Power Systems

Q.H. Wu · Z. Lu · T.Y. Ji

Protective Relaying of Power Systems Using Mathematical Morphology

Springer

Authors
Prof. Dr. Q.H. Wu
Dr. Z. Lu
T.Y. Ji

Dept. Electrical Engineering and Electronics
University of Liverpool
Brownlow Hill
Liverpool L69 3GJ
UK

ISSN 1612-1287
ISBN 978-1-84882-498-0 e-ISBN 978-1-84882-499-7
DOI 10.1007/978-1-84882-499-7
Springer Dordrecht Heidelberg London New York

British Library Cataloguing in Publication Data
A catalogue record for this book is available from the British Library

Library of Congress Control Number: 2009927244

Cover design: deblik, Berlin

Printed on acid-free paper

Springer is part of Springer Science+Business Media (www.springer.com)

To our families

Preface

The basic principle of protective relaying of power systems has not changed for more than half a century. Almost all power system protective relaying algorithms are dominated by integral transforms such as the Fourier transform and the wavelet transform. The integral transform can only provide an average attribute of the signals or their components. The accuracy of the attribute extraction is significantly sacrificed by the assumption of periodicity of the signals if the integral transform is applied to transient signals. It is also well known that the signals are liable to be contaminated by noise in the form of exponentially decaying DC offsets, high frequency transients, harmonic distortion, errors caused by non-linearity in the response of the sensors, and unwanted behaviour of power systems. This contamination is often provoked by fault conditions, just at the time when the protection relay is required to respond and trip the circuit breaker to limit damage caused by the fault.

On the other hand, as we know, in most protection relays, complex computation has to be undertaken within a sampling interval, no matter how small the interval, to calculate the coefficients relevant to the attributes of the signals by using the integral transform based on a window of samples, and to calculate the relaying algorithms, which are derived to represent the relationship between these coefficients and power system faults. If fast transients and high-order harmonics are to be addressed, extra computing power and facilities are required. Therefore, it can be seen that the current power system relaying algorithms suffer from many problems including accuracy, fast responses, noise, disturbance rejections and reliability.

To tackle the problems of distorted waveforms, disturbances and transient components of fault voltages and currents, identification of the shapes of complex waveforms is ideally required instead of the analysis of periodic characteristics, which is undertaken by the currently used integral transform to obtain the knowledge of distorted signals indirectly. However, there is currently no generic methodology available for designing a protection relay that is able to detect the shapes of signals, in particular for protective relaying purposes.

This book introduces mathematical morphology (MM) for the design and operation of power system relays. MM has been designated as a new branch of mathematics, which is totally different from the integral transform-based methods in basic principles, algorithmic operations and approach. When used for the extraction of waveform components, MM has the following merits in comparison with the integral transform methods: (1) The morphological operators have fast and simple calculations without using multiplication and division operations. (2) It is applicable to non-periodic transient signals and not restricted to periodic signals. (3) MM uses a much smaller sampling window for real-time signal processing, as it does not require the information of the full signal components. (4) It is able to accurately and reliably extract the signal components without causing any distortion, as it is a time-domain signal processing method that does not perform any integral transforms.

We wrote this book in the belief that MM would open an opportunity to develop more accurate, reliable and faster protective relaying algorithms, leading to a new generation of power system protection relays. Apart from being an introduction to basic and advanced MM operators, presented with their pseudo codes, this book presents a number of MM-based methods developed for power system protection relays. We hope that this book will be useful for those postgraduates, academic researchers and engineers working in the area of the design and development of power system protection relays.

We would like to thank Drs. M.H. Sedaaghi, P. Sun, D.J. Zhang and J.F. Zhang for their contributions, made during the period of their PhD studies undertaken at The University of Liverpool, to part of the achievements presented in this book. We would also like to thank Dr. S. Potts of AREVA for providing the model of current transformers for the study of inrush current identification and Mr. John Fitch of National Grid for supporting this work and providing useful discussions.

Special thanks go to Anthony Doyle (the Senior Editor), Nadja Kroke and Sorina Moosdorf (the Production Editor), and Claire Protherough and Simon Rees (Editorial Assistants) for their professional and efficient editorial work on this book. Our thanks are also extended to all colleagues in the Intelligence Engineering and Automation Research Group, The University of Liverpool, for all assistance provided, and who have not been specifically mentioned above.

The University of Liverpool, UK
December 2008

Prof. Dr. Q.H. Wu
Dr. Z. Lu
T.Y. Ji

Contents

List of Figures

List of Tables

Chapter 1
Introduction

1.1 Introduction and Definitions

In a large power system network comprising hundreds of complexly interacting elements, there always exists the possibility of disturbance and fault [117]. The operations of early fault identification and rapid equipment isolation are imperative to maintain the reliability and integrity of the system. The failure of insulation or breaking of conductors and accidents may cause extensive damage to the equipment and interruption in the power supply to the customers. To solve these problems, power system equipment must be protected adequately. Protection relays are designed and applied in power systems to protect equipment from excessive damage and to maintain system integrity and stability. The function of a protection relay in association with a circuit breaker is to remove the system element from which a fault has been developed as quickly as possible.

Fast and correct protection relays resume power supply swiftly, which can improve system reliability and safety. Besides speed, the other three important features of a satisfactory protection relay are reliability, selectivity and sensitivity.

- **Reliability** is a measure of the certainty of the protective relaying to function correctly when required to act. It has two aspects. Dependability: it must operate in the presence of a fault that is within its protection zone, and, security: it must refrain from operating unnecessarily for faults outside its protection zone or in the absence of a fault.
- **Sensitivity** is an ability of the protective relaying to identify an abnormal condition that exceeds a normal threshold value, and then initiates protective action to isolate the fault.
- **Selectivity** refers to the overall design of protective relaying wherein only those protection relays closest to a fault will operate to remove the fault component. This implies a grading of the protective relaying threshold, timing, or operating characteristics to obtain the desired selective operation. This restricts interruptions to only those components that are faulted.

- **Speed** is the ultimate goal of protective relaying, to remove a fault system element as quickly as possible. Therefore, the stability of the power system and the quality of power delivery are improved during a fault-caused transient condition.

1.2 Historical Background of Digital Protective Relaying Algorithms

The fuse was the first form of protection for power apparatus in early electrical power systems. It was constructed from a piece of copper designed to melt when excessive current flowed through it. As power systems developed rapidly, the fuse element was no longer able to provide adequate and satisfactory protection for power systems. Consequently, electro-mechanical technology-based relays replaced the fuse to protect power systems. However, these relays suffered from several associated disadvantages like long operating time due to the inertia of moving parts, the high burden on current and voltage transformers (CTs and VTs) and high maintenance cost, which made it necessary to use more sophisticated relaying techniques.

Static relays were the next generation of techniques used. The fabrication of static relays using solid-state components was an improvement over the electro-mechanical relays. Some of the drawbacks of the electro-mechanical relays were eliminated, but the static relays still suffered from other limitations such as sensitivity to temperature and voltage spikes, aging of components and damage due to overloading. The early designs of static relays were vacuum tube devices, which were introduced in the 1930s. These systems required considerable maintenance and were not as reliable as electro-mechanical relays. Hence, their applications were limited. In the early 1950s, transistor-based relays became available and immediately provided reliable protection for power systems, which caused the static relays to begin to disappear in quantity in the mid-1960s.

Associated with the development of integrated operational amplifiers, integrated relays are devised to improve the capabilities of relays. Integrated relays are similar to static relays in design and function, but use integrated circuits rather than discrete components. Integrated circuitry allows these relays to combine the three-phase current protection, metering, communication, control and monitoring capabilities.

The first digital relay was developed and reported by Rockefeller in the late 1960s [114]. Much literature reported digital relays shortly afterwards. Nowadays, the concept of digital relays is growing rapidly as digital computers become more powerful, reliable and cheaper. A typical digital relay has a measurement unit and a computation unit. The former is comprised of data acquisition, conversion and storage; the latter executes the protective relaying algorithm with the data provided by the measurement unit. The process of measurement, fault computation and decision making is completed within the time interval between consecutive data sampling instants. The inherent features of a digital relay, such as memory action and complex shaping of operational characteristics, lead to better performance and the integration of diagnostic functions make the relay much more reliable. The feature of advanced programmable functions maximises the flexibility and monitoring capabilities as well.

The micro-processor replaces most of the electronic circuitry, thereby maximising integration of advanced protection functions, control, monitoring, alarms, metering and communication into a single device. In comparison with electro-mechanical and static relays, digital relays have a higher immunity to electro-magnetic fields and transients and can be adapted to a wider range of operating temperatures. Besides the benefits of digital technology, properly designed digital relays are superior to conventional electro-mechanical and static relays in several ways. Some of the expected benefits of a digital protection scheme are the following [118]:

- **Economy** The cost of digital hardware is steadily decreasing. The cost of a digital relay is now less than that of a comparable analogue relay. Although the software development costs for a digital relay are high, compared with the development cost of any new device, they would be distributed over many similar devices.
- **Performance** The performance of the commercial digital relay is considered to be as good as the corresponding perfect analogue relay. A digital relay has some inherent features, such as memory action and complex shaping of operational characteristics, which lead to better performance.
- **Reliability** The realisation of a high level of diagnostic functions in a digital relay makes the relay much more reliable. For example, the digital relay can perform self-checks at regular intervals.
- **Flexibility** Different functions of a digital relay can be achieved by changing only the software program on the same hardware platform.
- **Background Tasks** The digital relay can work as a computer by taking over other tasks such as measuring voltages and currents, monitoring power flows, controlling the opening and closing of circuit breaks and switches, and providing backup for other devices.

1.3 Development of Protective Relaying Algorithms

The introduction of digital relays offers the possibility of developing new and complex relaying principles. Over the past 40 years of development, different types of algorithms have been developed for various digital protection applications.

1.3.1 Sinusoidal Waveform-Based Algorithms

The key assumption of sinusoidal waveform-based algorithms is that the fault current and voltage waveforms are sinusoidal. Many of these algorithms were developed at the beginning of computer technique applications. They include various algorithms, such as sample and first derivatives [98], first and second derivatives

[50, 115], two-sample techniques and three-sample techniques [50]. An experimental online system for the distance protection of a transmission line was installed in a 230 kV substation for field trial in 1972. It was developed jointly by Westinghouse and the Pacific Gas & Electric Company in the USA.

1.3.2 Digital Filtering Techniques

Other developments using digital filtering techniques, which occurred at about the same time, took into account the non-sinusoidal nature of the fault current and voltage waveforms. Researchers concentrated their efforts on extracting the fundamental components from distorted waveforms. There are two types of digital filtering techniques for protective relays, one based on signal parameter estimation, and the other performing power system parameter estimation.

Most present-day digital relays include some form of signal parameter estimation algorithm. The parameter estimation process is essentially a convolution filtering process of an electrical signal. Finite impulse response (FIR) filters are used in digital relays because its output signal can be related to the input signal by a convolution sum. The most common FIR filters in digital protective relays are Fourier-related filters. The Fourier transform (FT) was first applied to both current and voltage waveforms using a data window of less than one cycle of power frequency, and the Fourier coefficients extracted from analysis of fault current and voltage were then used to calculate the measured impedance [74, 101]. Similar methods such as Walsh functions [62] and Kalman filters [51, 52] were also suggested for protection functions. However, the FT-based filter is the most popular algorithm and has become standard in the industry. The computational cost of the FT-based filter is very low and good harmonic immunity can be achieved.

1.3.3 Least Squares-Based Methods

Curve fitting techniques have been used to fit fault current and voltage waveforms with an approximated function consisting of a fundamental sinusoid component, an exponentially decaying DC component and/or harmonics. The least squares-based methods [119] also have the common goal of extracting the fundamental components of current and voltage waveforms to calculate the impedance to the fault, or the comparison of current-based signals in digital differential protection.

The least squares-based methods are commonly used to estimate the parameters of a given model [43]. In the case of model identification of transmission lines, they are not influenced by the exponentially decaying DC component. The DC component is not treated as a noise in the signals, but is, in fact, treated as useful information in least squares-based methods. A secure trip signal can be issued after a short time delay based upon the impedance estimations provided by these algorithms for transmission line protection. The trip times are shorter than those achieved using signal model-based digital filtering algorithms.

1.3.4 Differential Equation Methods

Differential equation methods were developed based on the fact that all protected equipment can normally be represented by the mathematical differential equations of either first or second order. No special assumption has been made with regard to the content of the fault current and voltage waveforms. For example, the transmission line can be modelled by a series of lumped impedance and shunt capacitance, which results in a second-order differential equation to determine the resistance and the inductance of the line in such a way as to ignore any particular harmonic components [113, 130]. The idea was then developed for an experimental computer relaying system by General Electric in 1979 [22, 45].

1.3.5 Protection Based on Transient Signals

The utilisation of extra-high-voltage (EHV) transmission lines in modern power systems brings new and relatively difficult problems, which require fast fault clearance to improve the system stability. The development of transient-based protection systems for transmission lines started in the late 1970s [71]. Considerable efforts have been made to research the detection of fault-generated transients and have covered many aspects of power system protection.

Transient-based protection detects the occurrence of a fault by measuring high frequency transient current or voltage signals generated by the fault. The transient signals are directly extracted from the outputs of CTs or VTs, by using fast signal processing algorithms [14]. The fault can be identified accurately and rapidly, without considering the effect of fault path impedance, power swing or CT saturation. These transient-based protection techniques have been developed and applied in the fields of fault location, power line protection, power apparatus protection and circuit breaker auto re-closure, etc. [3, 24].

1.3.6 Adaptive Protection

The study of adaptive protection relays began in the 1980s [116]. With the development of computers, communication and artificial intelligence technology, realisation of adaptive protection relays have become possible. A function within a protection relay or system that automatically adjusts the operating characteristics of the relaying system in response to changing power system conditions is said to be adaptive.

Adaptive protection can automatically find the optimal condition depending on relay settings and alter the protection settings, functions and characteristics in realtime. It makes online changes in the digital relays possible by computing and transmitting new operating parameters from a central site or from an expanded communication system. Various adaptive protection systems have been developed since the 1980s, including rapid coordination calculation experiments, the use of synchronised phasor measurement [139], circuit breaker auto re-closure [6] and transformer protection, agent, decision trees, and wide area-based protection systems [35].

1.3.7 Artificial Neural Networks for Protective Relaying

Artificial neural networks (ANNs) are biologically inspired; that is, they are composed of elements that perform in a manner analogous to the most elementary functions of the biological neuron. These elements are then organised in a way that may be related to the anatomy of the human brain. From the engineering point of view, ANN, which is a technique for dealing with large scale patterns, may have any number of variables, so that it can detect and respond to all the parameters describing the process studied and determine its current state. Successful applications of ANN in engineering have demonstrated that this tool can be utilised as an alternative method to solve protective problems accurately and efficiently.

ANNs have demonstrated their potential in terms of handling non-linearity, generalisation, prediction capabilities and complex interactions of time varying voltage and current under a fault condition. Numerous ANN-based applications have been developed for power system protection, which include: directional protection, distance protection, differential protection, circuit breaker re-closure [4], fault location, phase selection [5] and unit protection [20]. These techniques improve the performance of protection relays greatly.

1.3.8 Wavelet Transform Methods

The wavelet transform (WT) is a new powerful mathematical tool introduced for digital signal processing. It has attracted great attention and has become more and more popular since the 1980s [96]. The WT procedure is to adopt a wavelet prototype function, called ‘analysing wavelet’ or ‘mother wavelet’. Temporal analysis is performed with a contracted, high-frequency version of the prototype wavelet, while frequency analysis is performed with a detailed, low-frequency version of the prototype wavelet. Compared to FT-based techniques, which rely on a single basis function, WT has a number of prototype basis functions for selection. An appropriate wavelet function is selected as a mother wavelet to perform analysis of a specific signal based on the shift and dilation of the selected wavelet.

The main attribution of WT is its ability to not only decompose a signal into its frequency components, but also provide a non-uniform division of the frequency domain, whereby it focuses on short time intervals for high frequency components and long intervals for low frequencies. WT provides a powerful time-frequency localisation property along with signal decomposition. Multi-resolution analysis of WT refers to the procedures for obtaining low-pass approximations and band-pass details from original signals. An approximation contains the general trend of original signal while a detail embodies the high-frequency contents. The original signal is divided into different scales of approximations and details through a succession of convolution processes. It is useful for the detection and analysis of signal features, especially for analysing transient phenomena such as those associated with faults or switching operations, since it has the ability to analyse a localised area of a signal and reveal aspects of data like break points or discontinuities. Thus, WT is useful in detecting the onset of a fault and has been widely applied for the study of protective relaying of power systems, including high impedance fault detection [104], fault-phase identification [148], travelling wave fault location [27] and directional transformer protection [120].

The development of protective relaying algorithms and the methodologies and technologies used to design protection relays are summarised in Fig. 1.1. It can be seen from the figure that the technologies have been updated from electromechanical components to micro-processors, the methodologies have remained unchanged for more than half a century; over this long period of time, integral transform-based methods have played a major role in relay design. These methods work principally based on assumption that the fault voltage and current are periodic signals and practically on measurement of fundamental frequency components of waveforms. As mentioned above, they are able to deal with transient signals and non-periodic signals, which are commonly met in power system fault scenarios.

1.4 Introduction of Mathematical Morphology to Protective Relaying of Power Systems

The basic principle of power system relay design and operation has not changed for more than half a century, even after the introduction of digital relays in the early 1980s to replace their analogue counterparts. All protection relays measure attributes of signals from sensors connected to the power system. The major attributes are those of the fundamental frequency of the power system, namely the magnitude and phase angle of voltages and currents, and the frequency of the mains. To detect distortions of the voltage or current signals, which may be caused by system disturbances, unbalanced operation and the influence of power electronics devices, harmonics analysis is generally applied. The concept of harmonics stems from the integral transform, such as FT and WT. These transforms are calculated using the sampled data of the signal, which is required to cover a certain period of time to reveal the periodic characteristics of the signal. Obviously, the integral transform-

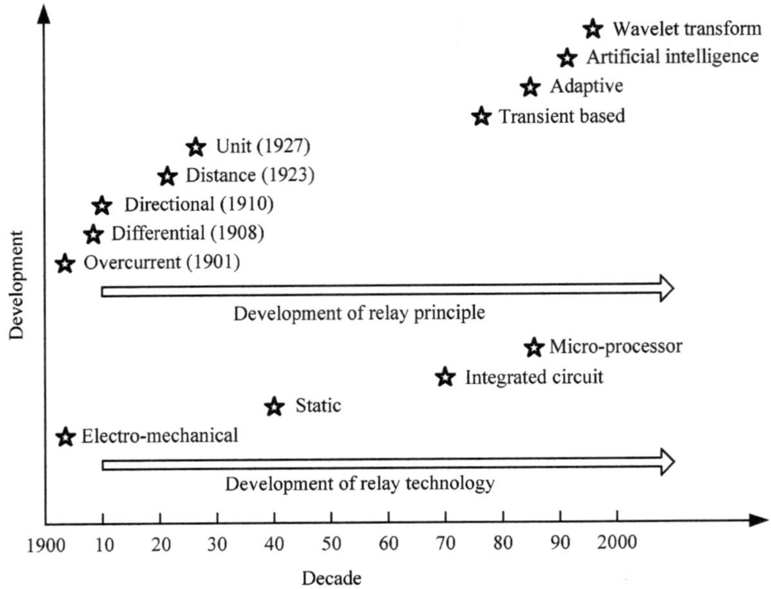

Fig. 1.1 The history of relay developments

based protective relaying algorithms can only provide an average attribute of the signals or their components. The accuracy of the attribute extraction is significantly sacrificed by the assumption of periodicity of the signals if the integral transform is applied to transient signals. It is also well known that the signals are liable to be contaminated by noise and distorted by disturbances in the form of exponentially decaying offsets, high frequency transients, errors caused by non-linearity in the response of the sensors, and unwanted behaviour of the power system plant. To tackle these disturbances, identification of the shapes of distorted waveforms is ideally required instead of the analysis of periodic characteristics, which is undertaken by the currently used integral transform to obtain the knowledge of distorted signals indirectly. However, there is currently no generic methodology available for designing a protection relay that is able to detect the shapes of signals. Concerning the implementation complexity, as we know, for most of the protection relays, two tasks have to be undertaken during a sampling interval, no matter how small the interval is: (1) to calculate the coefficients relevant to the attributes of the signals using the integral transform based on a window of samples; (2) to calculate the relaying algorithms that are derived to represent the relationship between these coefficients and power system faults. Therefore, it can be seen that the current relaying algorithms developed based on the conventional principle suffer from many problems including accuracy, response time, noise disturbance rejections and reliability.

This book is concerned with introducing mathematical morphology (MM) for protective relaying of power systems. MM was introduced in 1964 by Matheron [99] and Serra [125, 126], both researchers at the Paris School of Mines in Fontainebleau.

Their objective was to characterise physical or mechanical properties of certain materials, such as the permeability of porous media, by examining the geometrical structure [124]. In 1964, they both published their first reports concerning some of the methods that are part of MM now. Matheron rediscovered Minkowski algebra, which defined set addition (dilation) in 1903. In that period, Matheron had undertaken the study of permeability for porous media in relation with their geometry on a more theoretical level [100]. The notion of morphological opening and closing was introduced and associated with the convexity of a structuring element (SE), leading to granulometry. In January 1966, Serra finished the analysis of binary images using any SE based on straight lines along the three axes of the hexagonal grid. Later that year, Matheron and Serra coined the term 'mathematical morphology'. The Centre of MM, founded on April 20, 1968, is considered as the birthplace of MM, which is still an important centre for research in the field today.

Since its launching in 1964, MM grew rapidly in the 1970s on the theoretical level. The major contribution of Matheron's work during this period includes topological foundations, random sets, increasing mappings, convexity and several models of random sets. Based on iterative processing, binary thinning, skeletons, ultimate erosion, conditional bisectors and their geodesic framework were introduced. With the significant development of automated visual inspection, substantial developments in morphology were stimulated in the 1980s. The most important development of MM in that period was the setting of the method in the mathematical framework of complete lattices. The theory of the complete lattice provides a compact theoretical foundation for grey-scale morphological operations. The theory of morphological filtering was also presented in the 1980s. Some real-time applications of morphological filtering were developed, such as the Delft image processor (DIP). In the 1990s, the growth of MM was focused on its applications, which include robot vision, medical imaging, visual inspection, texture and scene analysis, etc. The theoretical works on the concept of connection, in combination with the connected filters, have made MM a remarkable powerful tool for segmentation [67, 134]. Slope transform was developed by endowing morphological operators with eigenfunctions and their related transfer functions. Furthermore, a relatively new approach of MM, soft morphology, was introduced by Koskinen [83]. The recent trends in soft morphology including the algorithms and implementations can be found in [47], [68] and [86]. Other extensions of MM include fuzzy morphologies, which attempt to apply fuzzy set theory to MM [17, 49, 127]. With over three decades of development, MM has become a powerful tool for geometrical shape analysis.

In the field of signal processing, MM prefers to depict the profile of signal waveforms in the time domain directly, in contrast to most filtering methodologies such as FT and WT, which place emphasis on the response in the frequency domain. The underlying basis of the morphological filtering technique is to process signals by a function, known as the SE or the structuring function generally. An SE slides through the signal as a moving window, inspects its interaction with the signal, and detects specific features in the neighbourhood of every point in the signal. A morphological filter is composed of various combinations of fundamental morphological operators, which can preserve or suppress the feature represented by SE and obtain a signal with only components of interest. When operating upon a signal of complicated shapes, morphological filters are capable of decomposing a signal into certain parts. These parts are separated from the background for identification or other purposes, and the main shape characteristics of the signal are preserved. Therefore, for a distorted signal, its underlying shapes can be identified, reconstructed and enhanced. Furthermore, the mathematical calculation involved in MM includes only addition, subtraction, maximum and minimum operations, without any multiplication or division. Hence, morphological operators calculate faster than traditional integral transforms when processing the same signal.

In contrast with the theory of linear signal processing, such as FT and WT, MM is concerned with the shape of a signal waveform in the time domain rather than the frequency domain. MM has been widely used in the areas of image processing, but only a few investigations have been attempted for signal processing. When used for the extraction of waveform components, MM has the following merits in comparison with the integral transform-based methods:

1. The morphological operators have fast and simple calculations without using multiplication and division operations.
2. It is applicable to non-periodic transient signals and not restricted to periodic signals.
3. MM uses a much smaller size of sampling window in real-time signal processing, as it does not require the information of the full signal components.
4. It is able to accurately and reliably extract the signal components without causing any distortion, as it is a time-domain signal processing method without performing any signal integral transforms.

With the help of the MM techniques, a new generation of protection relays, called morphological protection relays, are developed.

1.5 Contents of This Book

This book is organised as follows. Chapter 2 briefly introduces the mathematical background of MM, the basic morphological operators, such as dilation, erosion, opening and closing, and the advanced operators developed from the basic ones, which include a multi-resolution decomposition scheme and morphological wavelets. The pseudo codes of these operators are presented. This chapter provides the basic knowledge of MM operations for the development of MM-based protective relaying algorithms presented in the book.

Phasor measurement is commonly obtained using FT. However, when a fault occurs on transmission lines, fault currents always contain exponentially decaying DC offsets. The existence of the exponentially decaying DC offset degrades the accuracy of FT. Chapter 3 introduces MM-based methods to solve this problem. The methods can be used to remove the exponentially decaying DC offset from the fault currents before they are processed by FT, so that the performance of FT will not be jeopardised by the exponentially decaying DC offset and the accuracy of the phasor measurement is significantly improved.

Chapter 4 introduces two examples of the MM-based method used for protection of transmission lines. The first one is an adaptive distance relaying algorithm (ADRA), in which a morphological fault detector (MFD) is used to determine if a fault has occurred on the transmission line. The Euclidean norm of the MFD output is then calculated for fault phase selection and fault type classification. The fault location is then determined by an instantaneous circuit model, which is applicable to a transient fault process with respect to a specific type of fault scenario. Moreover, the tripping zone of ADRA is set in an adaptive manner, which guarantees the reliability of this protection scheme. The other example is concerned with the problem of mis-selection of fault phases during a fault occurring on a double circuit transmission line, due to the mutual coupling between the two circuits of the transmission line. An MM-based fault phase selection scheme is introduced to classify the fault types. In the scheme, the fault characteristics can be effectively extracted from the relevant currents by a morphological gradient wavelet (MGW), and then a simple criterion is introduced for fault phase selection.

Identification of the magnetising inrush current has been considered as a challenging problem for transformer protection. Chapter 5 deals with this problem. Two MM-based schemes are presented in this chapter. The first scheme detects the magnetising inrush current using a morphological decomposition scheme (MDS) to extract the singularity peculiar to the asymmetric inrush waveform, while attenuating other irrelevant features. The inrush current is then identified by quantifying the extracted features, using a set of simple criteria. The second is a multi-resolution decomposition scheme (MRDS) that builds on MDS but performs faster than the first scheme. By decomposing a current signal into meaningful levels, the scheme is able to discriminate between inrush and internal fault currents even in two extreme cases – an inrush with a low second harmonic component and an internal fault current with a high second harmonic component. The accuracy and reliability of this scheme qualify it as an alternative for transformer differential protection.

A major function of bus protection is provided by differential protection relays based on collected measurements via a number of current transformers (CTs). The CT saturation would significantly degrade the performance of a bus protection system. Chapter 6 introduces a morphological lifting scheme (MLS) to extract the features of the secondary current of CT to detect the CT saturation. A compensation algorithm is also developed to reconstruct healthy secondary currents for protection relays to function correctly and rapidly.

Chapters 7 and 8 are concerned with ultra-high-speed (UHS) protective relaying for EHV transmission lines. Chapter 7 concentrates on directional protection, which derives the direction discriminants of a fault occurring on a transmission line and determines the phase of the fault; while Chap. 8 focuses on positional protection, which determines the location of a fault by extracting the transient features. For the former, the multi-resolution morphological gradient algorithm (MMGA) is employed to extract the transient features of the wavefront of the travelling wave.

For the latter, which requires more accurate information of the fault, the noise is removed by MLS before the MMGA is applied to extract the wavefronts. Chapter 8 also introduces another scheme, the morphological undecimated wavelet (MUDW), for positional protection, which can deal with noise reduction and transient feature extraction at the same time.

Chapter 2
Mathematical Morphology

Abstract MM can be defined as a theory and technique for the analysis of spatial structures, based on set theory, integral geometry and lattice algebra. It is totally different from the methods that are based on integral transform, such as FT and WT, in basic principles, algorithmic operations and approach. In contrast to the theory of linear signal processing, such as FT/WT, MM is non-linear. Moreover, it is concerned with the shape of a signal waveform in the time domain rather than the frequency domain. The emphasis of this chapter is put on the methods and theory of MM, by introducing morphological operations for signal processing. After a brief introduction of the underlying mathematical concepts, a number of morphological operators and their features and functionalities are presented.

2.1 Introduction

MM was first introduced in 1964 by two French researchers, Matheron and Serra, who worked on problems in petrography and mineralogy [125]. They introduced a set formalism to analyse binary images, which allows the image objects to be processed by simple operations such as unions, intersections, complementation and translations. In 1975, a seminal book entitled *Random Sets and Integral Geometry* was published, which is considered to have laid down the foundations of MM [99]. Having been developed for several decades, MM has become a powerful tool for signal and image processing, especially for geometrical shape analysis.

In the past 40 years, a major research topic of MM has been digital image analysis. The development in terms of methods and applications has been equally boosted and there has always been a close interconnection between them. Mature and integrated methods have been formed, such as image filtering, image segmentation and classification, image measurements, pattern recognition, and texture analysis and synthesis, etc. As for the applications of MM, it includes visual inspection and quality control, optical character recognition and document processing, materials science, geosciences, and life sciences. An overview of the development of MM can be found in two milestone monographs by J. Serra in 1982 [125] and P. Soille in 2003 [131], respectively.

Considering its mathematical background, MM is defined on a complete lattice. Before giving the definition of a complete lattice, we need to understand the concept of a partly ordered set, which is also called a poset. A partly ordered set is a set in which a binary relation '$\leq$' is defined for certain pairs of elements. The binary relation '$\leq$' over a set $\mathscr{P}$ satisfies the following conditions for all elements $x, y, z \in \mathscr{P}$:

1. *Reflexive:* $\forall\, x, x \leq x$.
2. *Antisymmetry:* If $x \leq y$ and $y \leq x$, then $x = y$.
3. *Transitivity:* If $x \leq y$ and $y \leq z$, then $x \leq z$.

Given two partly ordered sets A and B and arbitrary elements a and x, the following definitions can be developed:

1. *Translation:* The translation of A by x, denoted by $(A)_x$, is defined as $(A)_x = \{a + x | a \in A\}$.
2. *Reflection:* The reflection of A, denoted by $\check{A}$, is defined as $\check{A} = \{-a | a \in A\}$. Reflection is also called transposition.
3. *Complement:* The complement of A, denoted by A^c, is defined as $A^c = \{x | x \notin A\}$.
4. *Difference:* The difference between two sets A and B, denoted by $A - B$, is defined as $A - B = \{x | x \in A, x \notin B\} = A \bigcap B^c$. Based on this operation, the complement of set A can also be defined as $A^c = \{x | x \in I - A\}$, where I is the universal set.

The partially ordered set formalises the intuitive concept of an ordering relation, which plays a key role in MM [131]. Aside from partial ordering, there also exists a total ordering relation. A totally ordered set has a strengthened relation of '$<$': for any two elements x and y, exactly one of $x < y$, $x = y$, $x > y$ is true. The property of transitivity on a totally ordered set becomes $x < y$ and $y < z$ implies $x < z$.

A poset, $(\mathscr{P}, \leq)$, is a lattice if any two elements of it, x and y, have a greatest lower bound (i.e. infimum), $x \wedge y$, and a least upper bound (i.e. supremum), $x \vee y$. A lattice, $\mathscr{J}$, is a complete lattice if each of its subsets has an infimum and a supremum in $\mathscr{J}$. A complete lattice satisfies the following properties: for subsets X, Y and Z,

1. *Commutativity:* $X \vee Y = Y \vee X, X \wedge Y = Y \wedge X$.
2. *Associativity:* $(X \vee Y) \vee Z = X \vee (Y \vee Z)$, $(X \wedge Y) \wedge Z = X \wedge (Y \wedge Z)$.

2.2 Basic Morphological Operators

2.2.1 Definitions for Binary Operations

2.2.1.1 Minkowski Addition and Subtraction

The main function of morphological operators is to extract relevant structures of a set. The extraction is usually done by the interaction between the set and another set, which is called structuring element (SE). The shape of the SE is pre-defined according to some a priori knowledge about the shape of the signal. There are two basic morphological operators, dilation and erosion, which form a pair of dual transforms. They are derived from Minkowski set theory. Therefore, we start this section from the introduction to Minkowski addition and subtraction.

Minkowski addition is a binary operation of two sets A and B in Euclidean space, named after Hermann Minkowski. It is denoted by $\overset{\mathrm{M}}{\oplus}$ and is defined as the result of adding every element of A to every element of B:

$$A \overset{\mathrm{M}}{\oplus} B = \{a+b | a \in A, b \in B\} = \bigcup_{b \in B} (A)_b. \tag{2.1}$$

It is assumed that $A \overset{\mathrm{M}}{\oplus} \{0\} = A$ and $A \overset{\mathrm{M}}{\oplus} \emptyset = \emptyset$. The dual operation is called Minkowski subtraction, denoted by $\overset{\mathrm{M}}{\ominus}$, and defined as:

$$A \overset{\mathrm{M}}{\ominus} B = \bigcap_{b \in B} (A)_b. \tag{2.2}$$

The following relation will be true for Minkowski subtraction:

$$A \overset{\mathrm{M}}{\ominus} B = \{x | (\check{B})_x \subseteq A\} = (A^{\mathrm{c}} \overset{\mathrm{M}}{\oplus} B)^{\mathrm{c}}. \tag{2.3}$$

Proof:

$$\begin{aligned}
\because \quad & z \in \left(\bigcap_{b \in B} (A)_b \right) \\
& \Rightarrow z \in (A)_b \\
& \Rightarrow z \in \{a+b | a \in A, b \in B\} \\
& \Rightarrow z \in \{x | x-b \in A, b \in B\} \\
\text{and } \because \quad & \{x | (\check{B})_x \subseteq A\} = \{x | x-b \in A, b \in B\} \\
\therefore \quad & A \overset{\mathrm{M}}{\ominus} B = \{x | (\check{B})_x \subseteq A\}.
\end{aligned}$$

$$\because A^c \overset{M}{\oplus} B = \bigcup_{b\in B}(I-A)_b = I - \bigcap_{b\in B}(A)_b$$
$$= \left(\bigcap_{b\in B}(A)_b\right)^c = \left(A \overset{M}{\ominus} B\right)^c$$
$$\therefore\ A \overset{M}{\ominus} B = (A^c \overset{M}{\oplus} B)^c.$$

Next, we will prove that Minkowski addition satisfies the properties of commutativity and associativity. Similar methods can be used to prove that Minkowski subtraction also satisfies these properties.

1. *Commutativity:* $A \overset{M}{\oplus} B = B \overset{M}{\oplus} A$.
   Proof:
$$A \overset{M}{\oplus} B = \{x|x = a+b, a\in A, b\in B\}$$
$$= \{x|x = b+a, b\in B, a\in A\} = B \overset{M}{\oplus} A.$$

2. *Associativity:* $(A \overset{M}{\oplus} B) \overset{M}{\oplus} C = A \overset{M}{\oplus} (B \overset{M}{\oplus} C)$.
   Proof:
$$A \overset{M}{\oplus} (B \overset{M}{\oplus} C) = \{a+(b+c)|a\in A, b+c\in B \overset{M}{\oplus} C\}$$
$$= \{a+b+c|a\in A, b\in B, c\in C\}$$
$$= \{(a+b)+c|a+b\in A \overset{M}{\oplus} B, c\in C\}$$
$$= (A \overset{M}{\oplus} B) \overset{M}{\oplus} C.$$

2.2.1.2 Binary Dilation and Erosion

The definition of dilation is similar to Minkowski addition. Replacing the operator $\overset{M}{\oplus}$ by $\oplus$ in (2.1) and use a reflected SE, $\check{B}$, we have

$$A\oplus B = \bigcup_{b\in\check{B}}(A)_b = \bigcup_{b\in B}(A)_{-b} = \bigcup_{b\in B}\{x|x = a-b, a\in A\}. \tag{2.4}$$

Erosion derives from Minkowski subtraction, but this time b comes from the reflection set of B. The definition of erosion is

$$A\ominus B = \bigcap_{b\in\check{B}}(A)_b = \bigcap_{b\in B}(A)_{-b} = \bigcap_{b\in B}\{x|x = a-b, a\in A\}. \tag{2.5}$$

Since a binary image is a digital image that has only two possible values for each pixel, it is very convenient to describe a binary image using the notion of a set. A binary image is often considered as a set I, while an object in it is considered as a subset $X\subseteq I$. Letting sets A and B represent two binary images in the above definitions, we have the operations of dilation and erosion for binary image processing.

The two images A and B function differently in image processing. Generally speaking, A is the image being processed, while B serves as an SE that slides as a probe across image A and interacts with each pixel of A. Obviously, the size of B should be much smaller than that of A. To have a clear view of this process, we give an example in Fig. 2.1 to show how dilation and erosion function between a binary image and an SE. Here, the origin of B is set at $(0,0)$.

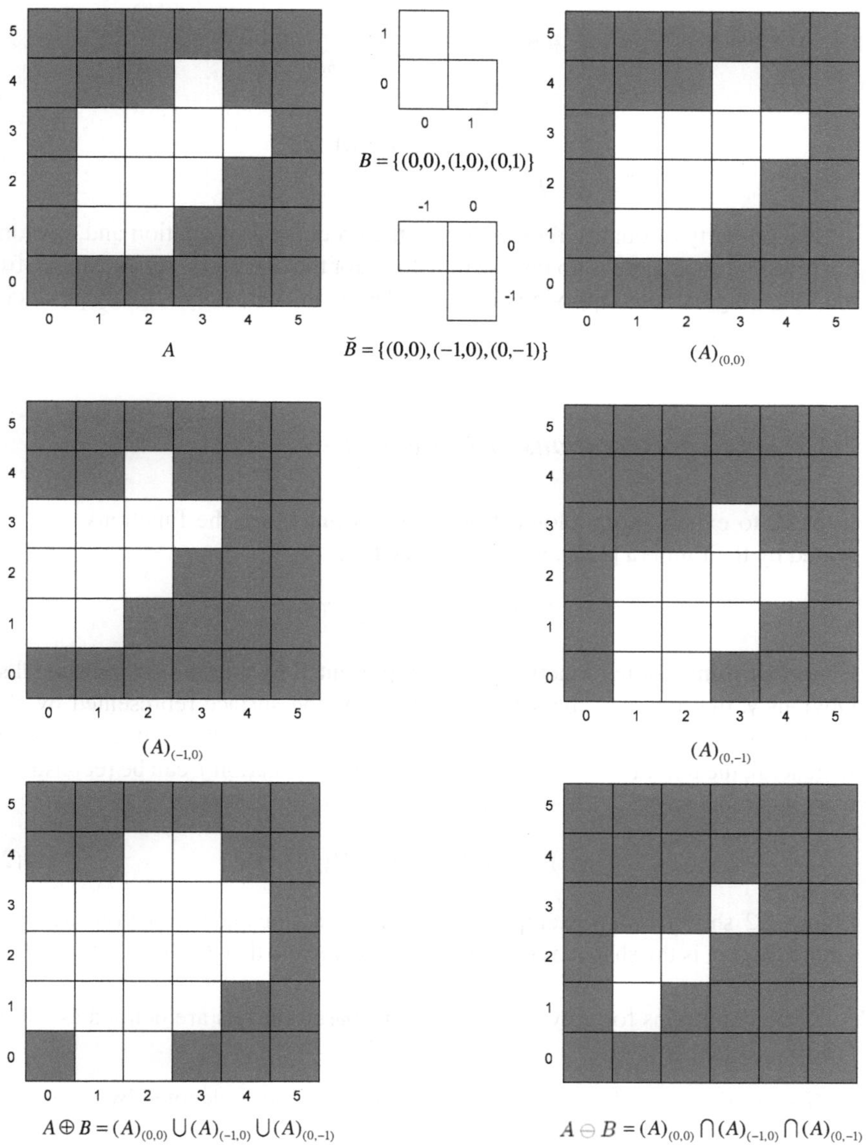

Fig. 2.1 Binary dilation and erosion of a binary image

Figure 2.1 illustrates an important property of dilation and erosion – duality, which means that applying dilation to A is equivalent to applying erosion to its complement A^c. This property can be expressed as:

$$A \oplus B = (A^c \ominus B)^c, \tag{2.6}$$

$$A \ominus B = (A^c \oplus B)^c. \tag{2.7}$$

Proof:

$$\begin{aligned} A^c \oplus B &= \bigcup_{b \in \check{B}} (I - A)_b = \bigcup_{b \in B} (I - A)_{-b} \\ &= I - \bigcap_{b \in B} (A)_{-b} \\ &= I - A \ominus B = (A \ominus B)^c \\ \Rightarrow\ A \ominus B &= (A^c \oplus B)^c. \end{aligned}$$

The property of duality illustrates that the processing of dilation and erosion is not reversible and there is no inverse transform for the operators. As we can see from the following sections, applying dilation and erosion alternately actually produces a pair of new operations.

2.2.2 Set Representations of Functions

In order to extend morphological operators to functions, the functions are represented by their umbra [132], which is defined as:

$$U(f) = \{(x,a) | a \leq f(x)\}. \tag{2.8}$$

Hence, a d-dimensional function $f(x)$ is represented by a $(d+1)$-dimensional set. Obviously, the umbra is the set of points below the surface represented by $f(x)$. After getting the umbra, binary morphological operators can be applied to the signal. In general, the umbra set extends to $a = -\infty$, and the function f can be reconstructed from its umbra since:

$$f(x) = \max\{a | (x,a) \in U(f)\},\ \forall x. \tag{2.9}$$

Figure 2.2 shows, as an example, the umbra of a sinusoidal function, where the umbra of $f(x)$ is the shaded region. We can easily show that $f \leq g \Leftrightarrow U(f) \subseteq U(g)$.

Some definitions for grey-scale operations based on sets are defined as follows [140]:

1. *Grey-scale union:* The union of two functions f and g, denoted by $f \vee g$, is defined as:

$$(f \vee g)(x) = f(x) \vee g(x). \tag{2.10}$$

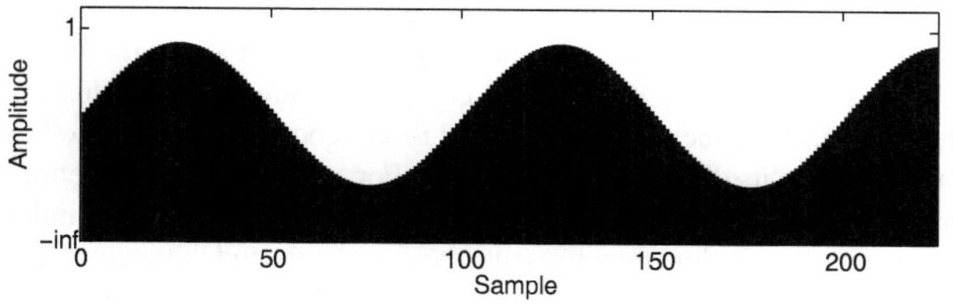

Fig. 2.2 Umbrae $U(f)$ of a sinusoidal function f

There is a one-to-one correspondence between the union of functions and the set union:

$$U(f \vee g) = U(f) \cup U(g). \tag{2.11}$$

2. *Grey-scale intersection:* The intersection of two functions f and g, denoted by $f \wedge g$, is defined as:

$$(f \wedge g)(x) = f(x) \wedge g(x). \tag{2.12}$$

A similar one-to-one correspondence exists for the function and the set intersection:

$$U(f \wedge g) = U(f) \cap U(g). \tag{2.13}$$

3. *Grey-scale transpose:* The transpose $\check{f}$ of a function f is defined as:

$$\check{f}(x) = f(-x). \tag{2.14}$$

4. *Grey-scale complement:* The complement f^{c} of a function f is defined as:

$$f^{\mathrm{c}}(x) = -f(x). \tag{2.15}$$

Notice that $f \vee f^{\mathrm{c}} = |f|$ and $f \wedge f^{\mathrm{c}} = -|f|$, whereas for sets, we have $A \cup A^{\mathrm{c}} = I$ and $A \cap A^{\mathrm{c}} = \emptyset$. Figure 2.3 shows the above properties.

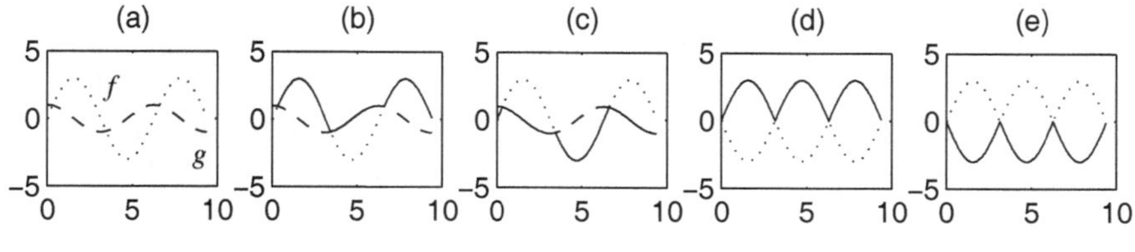

Fig. 2.3 The grey-scale operations. **a** Two functions f (*dotted line*) and g (*dashed line*). **b** $f \vee g$ (*solid line*). **c** $f \wedge g$ (*solid line*). **d** $f \vee f^{\mathrm{c}}$ (*solid line*). **e** $f \wedge f^{\mathrm{c}}$ (*solid line*)

For an eight-bit grey-scale digital image, each pixel of which can have $2^8 = 256$ possible values to carry the full and only information about its intensity, the image can be represented as a set whose components are in $\mathbb{Z}^3$. In this case, two components of each element of the set refer to the coordinates of a pixel, and the third corresponds to its discrete intensity value. For a signal, the set is defined in $\mathbb{Z}^2$ with each element corresponding to a sample of the digitised signal. Similarly, the first component of each element represents the coordinate and the second represents its value. Sets in a higher dimensional space can contain other attributes, such as the colour information of an image.

2.2.3 Grey-Scale Dilation and Erosion

In order to use MM in signal processing where most signals are not binary, morphological operators should be extended to a grey-scale level. Instead of performing dilation and erosion by union and intersection as in the binary case, they are performed by algebraic addition and subtraction in the grey-scale case. Let f denote a signal and g denote an SE, and the length of g be considerably shorter than that of f [122]. Dilation and erosion are defined as follows:

$$f \oplus g(x) = \max_{s}\{f(x+s)+g(s) | (x+s) \in \mathscr{D}_f, s \in \mathscr{D}_g\}, \tag{2.16}$$

$$f \ominus g(x) = \min_{s}\{f(x+s)-g(s) | (x+s) \in \mathscr{D}_f, s \in \mathscr{D}_g\}, \tag{2.17}$$

where $\mathscr{D}_f$, $\mathscr{D}_g$ are the definition domains of f and g, respectively. For example, suppose the SE, g, has a length of five samples with its origin in the middle. In this case, the domain of g is given by $\mathscr{D}_g = \{-2,-1,0,1,2\}$. The dilation and erosion of f by g are therefore calculated from

$$\begin{aligned} f \oplus g(x) = \max\{&f(x-2)+g(-2), f(x-1)+g(-1), f(x)+g(0),\\ &f(x+1)+g(1), f(x+2)+g(2)\}, \end{aligned}$$

and

$$\begin{aligned} f \ominus g(x) = \min\{&f(x-2)-g(-2), f(x-1)-g(-1), f(x)-g(0),\\ &f(x+1)-g(1), f(x+2)-g(2)\}, \end{aligned}$$

respectively. Intuitively, dilation can be imagined as swelling or expanding, while erosion can be thought of as a shrinking procedure. The pseudo code of dilation is given in Table 2.1. Likewise, the pseudo code of erosion can be given in a similar way by substituting 'maximum' with 'minimum', as shown in Table 2.2.

As explained previously, the SE is a small set used to probe the signal under study. A simple case is that the SE has the form of $g(s) \equiv 0$, $s \in \mathscr{D}_g$, which is referred to as a 'flat SE'. Hence, definitions of dilation and erosion degrade to:

Table 2.1 The pseudo code of dilation

```
Determine the SE, including its definition domain and the value of each element. Suppose m ≤ s ≤ n;
for (each sample of the signal f(x))
  for (m ≤ s ≤ n)
    Calculate ω(s − m + 1) = f(x + s) + g(s);
  end
  Return the maximum element of ω and f ⊕ g(x) = max{ω};
end
```

Table 2.2 The pseudo code of erosion

```
Determine the SE, including its definition domain and the value of each element. Suppose m ≤ s ≤ n;
for (each sample of the signal f(x))
  for (m ≤ s ≤ n)
    Calculate ϖ(s − m + 1) = f(x + s) − g(s);
  end
  Return the minimum element of ϖ and f ⊖ g(x) = min{ϖ};
end
```

$$f \oplus g(x) = \max_s\{f(x+s) \mid (x+s) \in \mathscr{D}_f, s \in \mathscr{D}_g\}, \tag{2.18}$$

$$f \ominus g(x) = \min_s\{f(x+s) \mid (x+s) \in \mathscr{D}_f, s \in \mathscr{D}_g\}. \tag{2.19}$$

The function of g is to indicate which samples are involved when processing the current sample. For a binary signal, the SE, g, must be flat. The dilation and erosion of a one-dimensional signal are illustrated in Figs. 2.4a and b, respectively. Both operations use a flat SE of length 3: $g(-1) = g(0) = g(1) = 0$.

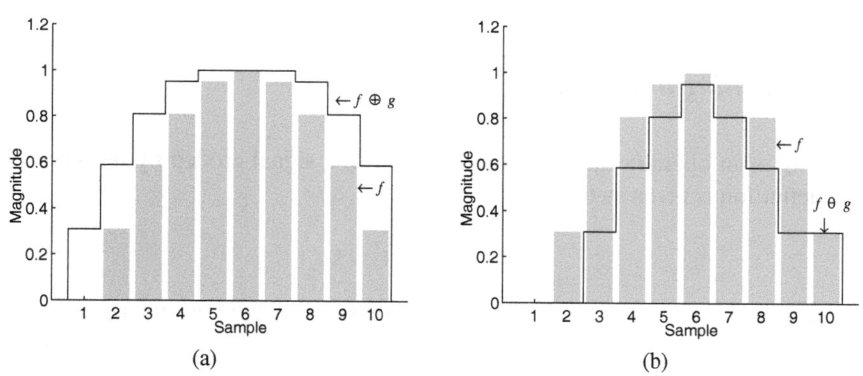

Fig. 2.4 Grey-scale dilation and erosion of a one-dimensional signal. **a** Dilation. **b** Erosion

Figure 2.4 demonstrates another property of dilation and erosion, i.e. that they are increasing transforms. The property can be expressed as follows. For two signals,

f_1 and f_2, and an arbitrary SE, g, we have:

$$f_1 \le f_2 \Rightarrow \begin{cases} f_1 \oplus g \le f_2 \oplus g \\ f_1 \ominus g \le f_2 \ominus g \end{cases}. \tag{2.20}$$

The ordering relation between dilation and erosion can be expressed as the erosion of a signal by an SE being less than or equal to its dilation by the same SE: $f \ominus g \le f \oplus g$. If the SE contains its origin, which means processing a sample of the signal within a window that contains the sample, the following ordering exists:

$$f \ominus g \le f \le f \oplus g. \tag{2.21}$$

2.3 Morphological Filters

2.3.1 Definitions of Morphological Filters

Morphological filters are non-linear signal transforms that locally modify the geometrical features of signals or image objects. The idempotence and increasing properties are necessary and sufficient conditions for a transform, ψ, to be a morphological filter:

ψ is a morphological filter $\Leftrightarrow$ ψ is increasing and idempotent.

The property of idempotence implies that applying a morphological filter twice to a signal is equivalent to applying it only once:

ψ is idempotent $\Leftrightarrow$ $\psi\psi = \psi$.

The increasing property ensures that the ordering relation on signals is preserved after being filtered if the same SE is employed.

2.3.2 Opening and Closing

Opening is an operator that performs dilation on a signal eroded by the same SE. The definition is given as follows:

$$f \circ g = (f \ominus g) \oplus g, \tag{2.22}$$

where f is the signal, g is the SE, and $\circ$ denotes the opening operator. Opening can recover most structures lost by erosion, except for those completely erased by erosion. Closing, on the other hand, can be defined by its duality as:

$$f \bullet g = (f \oplus g) \ominus g. \tag{2.23}$$

Usually, opening and closing are also denoted by operators γ and ϕ, respectively. The pseudo code of opening is given in Table 2.3, and the pseudo code of closing can be given likewise, as shown in Table 2.4. The results of performing opening and closing on the signal used in the previous section by the same SE are illustrated in Fig. 2.5.

Table 2.3 The pseudo code of opening

```
Determine the SE, including its definition domain and the value of each element. Suppose m ≤ s ≤ n;
for (each sample of the signal f(x))
  for (m ≤ s ≤ n)
    Calculate ϖ(s − m + 1) = f(x + s) − g(s);
  end
  Return the minimum element of ϖ and ε(x) = min{ϖ};
end
for (each sample of the signal ε(x))
  for (m ≤ s ≤ n)
    Calculate ω(s − m + 1) = ε(x + s) + g(s);
  end
  Return the maximum element of ω and γ(x) = max{ω};
end
```

Table 2.4 The pseudo code of closing

```
Determine the SE, including its definition domain and the value of each element. Suppose m ≤ s ≤ n;
for (each sample of the signal f(x))
  for (m ≤ s ≤ n)
    Calculate ω(s − m + 1) = f(x + s) + g(s);
  end
  Return the maximum element of ω and δ(x) = max{ω};
end
for (each sample of the signal δ(x))
  for (m ≤ s ≤ n)
    Calculate ϖ(s − m + 1) = δ(x + s) − g(s);
  end
  Return the minimum element of ϖ and φ(x) = min{ϖ};
end
```

Morphological opening and closing are both increasing transforms:

$$f_1 \leq f_2 \Rightarrow \begin{cases} \gamma(f_1) \leq \gamma(f_2) \\ \phi(f_1) \leq \phi(f_2) \end{cases}. \tag{2.24}$$

Moreover, successive applications of openings or closings do not further modify the signal, which means that they are both idempotent transforms:

$$\begin{aligned} \gamma\gamma &= \gamma \\ \phi\phi &= \phi \end{aligned}. \tag{2.25}$$

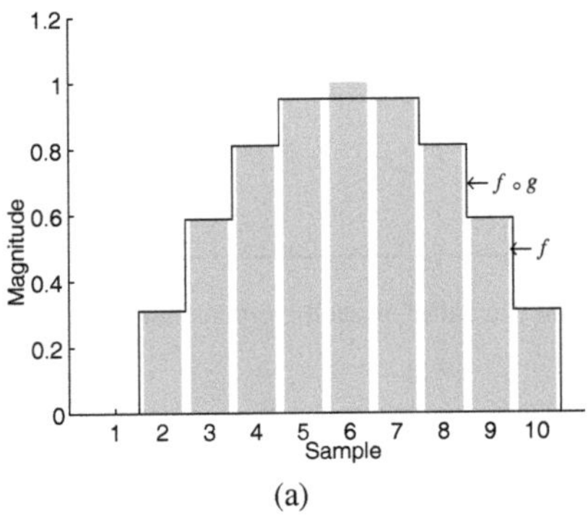

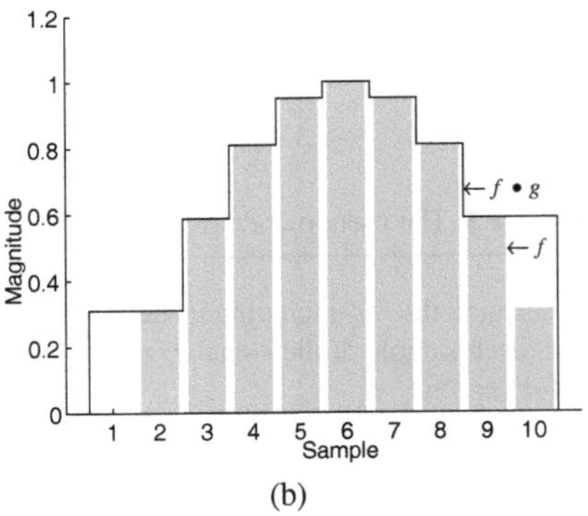

Fig. 2.5 Grey-scale opening and closing of a one-dimensional signal. **a** Opening. **b** Closing

Apparently, opening and closing fulfill the conditions of morphological filters.

Opening and closing are also a pair of dual transforms:

$$A \circ B = (A^c \bullet B)^c, \tag{2.26}$$

$$A \bullet B = (A^c \circ B)^c. \tag{2.27}$$

Proof:

$$\begin{aligned} (A \circ B)^c &= (A \ominus B \oplus B)^c = (A \ominus B)^c \ominus B \\ &= A^c \oplus B \ominus B = A^c \bullet B \\ \Rightarrow \quad A \circ B &= (A^c \bullet B)^c. \end{aligned}$$

A useful operator called top-hap transform can be derived from opening, which is defined as:

$$T_{TH}(f) = f - f \circ g, \tag{2.28}$$

where $T_{TH}(f)$ denotes the top-hap transform of signal f. This operator is useful for enhancing the detail in the presence of shading [26].

2.3.3 Alternating Sequential Filters

Opening and closing are the basic morphological filters and new filters can be designed from their sequential combinations, such as an opening followed by a closing or vice versa. In fact, all the following combinations are filters: $\gamma\phi$, $\phi\gamma$, $\gamma\phi\gamma$ and $\phi\gamma\phi$. Moreover, for these new filters, the ordering relations of

$$\gamma \le \gamma\phi\gamma \le \begin{matrix} \gamma\phi \\ \phi\gamma \end{matrix} \le \phi\gamma\phi \le \phi \tag{2.29}$$

are always satisfied [131]. The pair of dual filters $\gamma\phi$ and $\phi\gamma$ are called opening-closing and closing-opening filters, and they have almost the same filtering effects. Therefore, in practice, usually only one of them is employed.

In some applications, such as when the objects under processing are over a wide range of sizes, there is a need to alternatively use openings and closings with an SE of an increasing size. This sequential application of the basic operators is called an alternating sequential filter. Since the four types of sequential combinations of opening and closing are all morphological filters, four alternating sequential filters can be developed hereafter. Let γ_i and ϕ_i be a pair of dual operators with an SE of size i. Suppose the size of SE increases from i to j. Therefore, the four types of alternating sequential filters are given as:

$$f_{\text{aoc}} = (\gamma_j\phi_j)\cdots(\gamma_i\phi_i), \tag{2.30}$$
$$f_{\text{aco}} = (\phi_j\gamma_j)\cdots(\phi_i\gamma_i), \tag{2.31}$$
$$f_{\text{aoco}} = (\gamma_j\phi_j\gamma_j)\cdots(\gamma_i\phi_i\gamma_i), \tag{2.32}$$
$$f_{\text{acoc}} = (\phi_j\gamma_j\phi_j)\cdots(\phi_i\gamma_i\phi_i). \tag{2.33}$$

2.4 The Lifting Scheme and Morphological Wavelets

2.4.1 The Multi-resolution Decomposition Scheme

The multi-resolution decomposition scheme provides convenient and effective approaches for signal processing [97]. The core idea is to remove high frequencies in a signal, thus to obtain a reduced-scale version of the original signal. Therefore, by repeating this procedure, a collection of coarse signals at different levels are produced. Meanwhile, a collection of detail signals can be constructed by subtracting the coarse signal at level $i+1$ from the coarse signal at level i.

Let $J \subseteq \mathbb{N}$ be an index set indicating the levels in a multi-resolution decomposition scheme; J can be finite or infinite. Therefore, for a domain $V_j (j \in J)$, signals of level j belong to V_j. The analysis operator $\psi^{\uparrow}$ decomposes a signal in the direction of increasing, i.e. $\psi_j^{\uparrow} : V_j \to V_{j+1}$. On the other hand, the synthesis operator $\psi^{\downarrow}$ proceeds in the direction of decreasing, i.e. $\psi_j^{\downarrow} : V_{j+1} \to V_j$. In other words, the analysis operator $\psi^{\uparrow}$ maps a signal to a higher level and reduces information, while $\psi^{\downarrow}$ maps it to a lower level.

Composing analysis operators successively, a signal at any level i can be transferred to a higher level j. The composed analysis operator is denoted as $\psi_{i,j}^{\uparrow}$ and is defined to be:

$$\psi_{i,j}^{\uparrow} = \psi_{j-1}^{\uparrow}\psi_{j-2}^{\uparrow}\cdots\psi_i^{\uparrow}, \quad j > i. \tag{2.34}$$

It maps an element in V_i to an element in V_j. It should be noted that here the operations are carried from right to left. Likewise, the composed synthesis operator:

$$\psi_{j,i}^{\downarrow} = \psi_i^{\downarrow} \psi_{i+1}^{\downarrow} \cdots \psi_{j-1}^{\downarrow}, \quad j > i \tag{2.35}$$

takes the signal back from level j to level i. Finally, the composition operator that takes a signal from level i to level j and then back to level i is defined as:

$$\hat{\psi}_{i,j} = \psi_{j,i}^{\downarrow} \psi_{i,j}^{\uparrow}, \quad j > i, \tag{2.36}$$

where $\hat{\psi}_{i,j}$ is viewed as an approximation operator because analysis operator $\psi_j^{\uparrow}$ reduces signal information and, in general, the synthesis operator $\psi_j^{\downarrow}$ cannot compensate the information lost in the analysing procedure.

Analysis and synthesis operators play an important role in the construction of a decomposition scheme. If there is no information lost during the analysis and synthesis procedure, then this decomposition scheme would be perfect. The lifting scheme [136, 137] and morphological wavelets [55, 60] discussed in the next sections are designed to not modify the decomposition when the analysis and synthesis steps repeat.

2.4.2 The Lifting Scheme

The lifting scheme or simple lifting was originally developed as an alternative way to construct the wavelets used in (the first generation) WT, which leads to the so-called second generation WT [136, 137]. The second generation WT not only preserves the good features of the the first generation WT, namely time-frequency localisation and fast algorithms, but can also extend beyond simple geometries. As an alternate implementation of classical WTs, the lifting scheme provides an entirely spatial-domain interpretation of the transform, as opposed to the traditional frequency domain-based constructions. The feature of local spatial interpretation enables us to adapt the transform to the data, thereby introducing non-linearities in the process of multi-scale transforms. Compared with FTs using the same filter all the time and wavelets being translation and dilation of one given function, lifting adapts local data irregularities in the transform process.

The basic idea of the lifting scheme is very simple: the redundancy is removed by using the correlation in the data. To reach this goal, a signal x_k is first split into two sets – the even set $\{x_k | k \text{ is even}\}$ and the odd set $\{x_k | k \text{ is odd}\}$. Then, the odd set is predicted from the even set. Finally, the prediction error, which is the difference between the odd-indexed sample and its prediction, is used to update the even-indexed data. The three steps are noted as split, predict and update, respectively, as shown in Fig. 2.6.

1. *Split stage:* Let $x(n)$ represent a discrete signal. Firstly, $x(n)$ can be split into even and odd components, $x_{\mathrm{e}}(n)$ and $x_{\mathrm{o}}(n)$, respectively, where

$$\begin{aligned} x_{\mathrm{e}}(n) &= x(2n), \\ x_{\mathrm{o}}(n) &= x(2n+1). \end{aligned} \tag{2.37}$$

If the signal $x(n)$ corresponds to the sample of a smooth and slowly varying function, then the components of $x_e(n)$ and $x_o(n)$ are highly correlated. Therefore, it should be possible to accurately predict each odd polyphase coefficient from the nearby even polyphase coefficients.

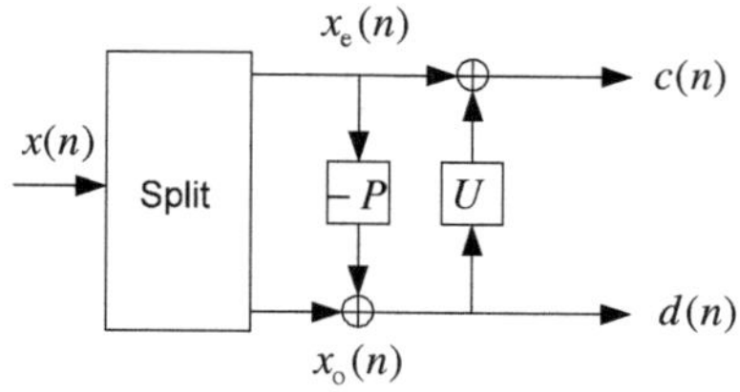

Fig. 2.6 The typical lifting steps (analysis): split, predict (P) and update (U). $c(n)$ and $d(n)$ are the approximation and detail signals, respectively

2. *Predict stage:* In the interpolating formulation of lifting, we can predict the odd polyphase coefficients $x_o(n)$ from the neighbouring even coefficients $x_e(n)$. Normally, the predictor for each $x_o(n)$ is a linear combination of neighbouring even coefficients:

$$P(x_e(n)) = \sum_l p_l x_e(n+l), \tag{2.38}$$

where the value of l lies on how many even coefficients are used for the prediction. If N ($N = 2D$) coefficients are used in symmetry, then:

$$-D+1 \le l \le D. \tag{2.39}$$

The prediction calculation is conducted within a data window of a fixed length. For each $x_o(n)$, its neighbouring even coefficients at both the left and right sides are involved in the calculation. As the window slides through the signal, the prediction for all odd coefficients is performed. If $n+l$ exceeds the range of the signal, the signal will be extended with zero samples, and this process is called zero padding.

A new representation of $x(n)$ may be obtained by replacing $x_o(n)$ with the prediction residual, $d(n)$:

$$d(n) = x_o(n) - P(x_e(n)). \tag{2.40}$$

The prediction residual will be small if the underlying signal is locally smooth. Furthermore, the same information of the original signal $x(n)$ is preserved in the residuals $d(n)$ since the odd polyphase coefficients $x_o(n)$ can be recovered by noting that

$$x_o(n) = d(n) + P(x_e(n)). \tag{2.41}$$

Actually, the prediction procedure is equivalent to applying a high pass filter to the original signal $x(n)$.

3. *Update stage:* The last lifting step transforms the even polyphase coefficients, $x_e(n)$, into a low pass filtered and sub-sampled version of $x(n)$ by updating $x_e(n)$ with a linear combination of the prediction residuals, $d(n)$, where $x_e(n)$ is replaced with
$$c(n) = x_e(n) + U(d(n)), \tag{2.42}$$
and normally $U(d)$ is a linear combination of the values of d:
$$U(d(n)) = \sum_l u_l d(n+l). \tag{2.43}$$
No information is lost in the update step. Assuming the same U is chosen for synthesis, given $d(n)$ and $c(n)$, we have
$$x_e(n) = c(n) - U(d(n)). \tag{2.44}$$

Following the description in WT, $d(n)$ is the detail signal obtained by the lifting scheme, while $c(n)$ is the approximation signal. Like WT, the aforementioned three stages are carried out on $c(n)$ recursively. The pseudo code of the lifting scheme is given in Table 2.5.

Table 2.5 The pseudo code of the lifting scheme

Denote the input signal $x(n)$ by $s_0(n)$. Set $k := 0$;
while (the end condition is not met)
  Split $s_{k,2l}(n) = s_k(2n)$;
  $s_{k,2l+1}(n) = s_k(2n+1)$;
  Predict Obtain the detail signal $d_{k+1,l}$:
  $d_{k+1,l}(n) = s_{k,2l+1}(n) - P(s_{k,2l}(n))$;
  Update Obtain the approximation signal $s_{k+1,l}$:
  $s_{k+1,l}(n) = s_{k,2l}(n) + U(d_{k+1,l}(n))$;
  $k := k+1$;
end

The inverse lifting stage is shown in Fig. 2.7. It corresponds to a critically sampled perfect reconstruction filter bank with c and d at half a rate.

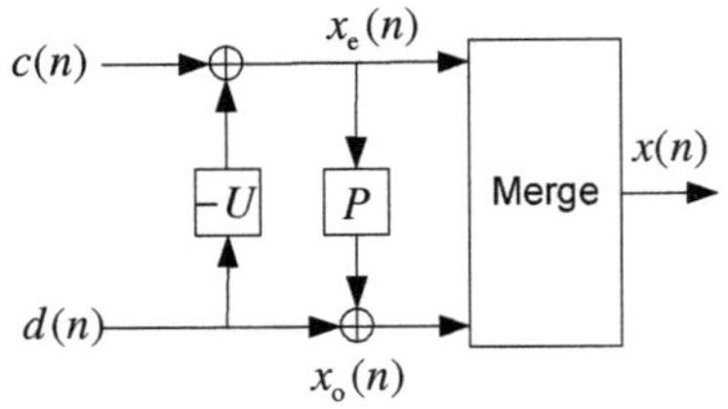

Fig. 2.7 The typical inverse lifting steps (synthesis): update (U), predict (P) and merge

As an example, the lifting scheme of the Harr wavelet and a family of Deslauriers–Dubuc wavelets are demonstrated here. The definitions of the first level are given as follows:

Harr:

$$\begin{aligned} d_{1,l} &= s_{0,2l+1} - s_{0,2l} \\ s_{1,l} &= s_{0,2l} + d_{1,l}/2 \end{aligned} . \quad (2.45)$$

Deslauriers–Dubuc:

$$\begin{aligned} (2,2): d_{1,l} &= s_{0,2l+1} - 1/2(s_{0,2l} + s_{0,2l+2}) \\ s_{1,l} &= s_{0,2l} + 1/4(d_{1,l-1} + d_{1,l}) \end{aligned} , \quad (2.46)$$

$$\begin{aligned} (4,2): d_{1,l} &= s_{0,2l+1} - [9/16(s_{0,2l} + s_{0,2l+2}) \\ &\quad -1/16(s_{0,2l-2} + s_{0,2l+4})] \\ s_{1,l} &= s_{0,2l} + 1/4(d_{1,l-1} + d_{1,l}) \end{aligned} , \quad (2.47)$$

$$\begin{aligned} (6,2): d_{1,l} &= s_{0,2l+1} - [75/128(s_{0,2l} + s_{0,2l+2}) \\ &\quad -25/256(s_{0,2l-2} + s_{0,2l+4}) \\ &\quad +3/256(s_{0,2l-4} + s_{0,2l+6})] \\ s_{1,l} &= s_{0,2l} + 1/4(d_{1-1,l} + d_{1,l}). \end{aligned} . \quad (2.48)$$

2.4.3 Morphological Wavelets

Most modern multi-resolution decomposition schemes are based on the pyramid and WT theories, using the convolution and time-frequency domain transformations. However, the linear filtering approaches to multi-resolution decomposition have not been theoretically justified. In particular, the operators used for generating various levels of signal components in a pyramid must crucially depend on an application. Therefore, in recent years, a number of researchers have proposed non-linear multi-resolution signal decomposition schemes based on morphological operators. The fundamental theories of the morphological pyramid [55] and the morphological wavelet [60] have built a framework for non-linear pyramids, filter banks and wavelets construction. They inherit the multi-dimension and multi-level analysis of wavelet and pyramid, while they do not require time-frequency domain analysis.

The morphological wavelet is a non-linear multi-resolution signal decomposition scheme [60]. It differs from ordinary multi-resolution decomposition schemes in that it engages two analysis operators and one synthesis operator. A formal definition of the morphological wavelet is presented as follows. Assume that there exist sets V_j and W_j. V_j is referred to as the signal space at level j and W_j the detail space at level j. The two analysis operators together decompose a signal in the direction of increasing j. The signal analysis operator, $\psi_j^{\uparrow}$, maps a signal from V_j to V_{j+1}, i.e. $\psi_j^{\uparrow} : V_j \rightarrow V_{j+1}$, while the detail analysis operator, $w_j^{\uparrow}$, maps it from V_j to W_{j+1}, i.e. $w_j^{\uparrow} : V_j \rightarrow W_{j+1}$. On the other hand, a synthesis operator proceeds in the direction of decreasing j, denoted as $\Psi_j^{\downarrow} : V_{j+1} \times W_{j+1} \rightarrow V_j$.

In order to yield a complete signal representation, the mappings $(\psi_j^{\uparrow}, \omega_j^{\uparrow}) : V_j \rightarrow V_{j+1} \times W_{j+1}$ and $\Psi_j^{\downarrow} : V_{j+1} \times W_{j+1} \rightarrow V_j$ should be inverses of each other, which means that the following condition:

$$\Psi_j^{\downarrow}(\psi_j^{\uparrow}(x), \omega_j^{\uparrow}(x)) = x,\ x \in V_j \tag{2.49}$$

should be fulfilled. This is called the perfect reconstruction condition, and

$$\begin{cases} \psi_j^{\uparrow}(\Psi_j^{\downarrow}(x,y)) = x,\ x \in V_{j+1}, y \in W_{j+1}, \\ \omega_j^{\uparrow}(\Psi_j^{\downarrow}(x,y)) = y,\ x \in V_{j+1}, y \in W_{j+1}, \end{cases} \tag{2.50}$$

where x is called the approximation signal and y is the detail signal. Therefore, decomposing an input signal $x_0 \in V_0$ with the following recursive analysis scheme:

$$x_0 \rightarrow \{x_1, y_1\} \rightarrow \{x_2, y_2, y_1\} \rightarrow \cdots \rightarrow \{x_j, y_j, y_{j-1}, \cdots, y_1\} \rightarrow \cdots \tag{2.51}$$

where

$$x_{j+1} = \psi_j^{\uparrow}(x_j) \in V_{j+1}, \tag{2.52}$$

$$y_{j+1} = w_j^{\uparrow}(x_j) \in W_{j+1}. \tag{2.53}$$

x_0 can be exactly reconstructed from x_j and y_j, y_{j-1}, ..., y_1 by means of the following recursive synthesis scheme:

$$x_{j-1} = \Psi_{j-1}^{\downarrow}(x_j, y_j). \tag{2.54}$$

This signal representation scheme is referred to as the morphological wavelet decomposition scheme and is illustrated in Fig. 2.8.

2.4.3.1 The Morphological Haar Wavelet

To make the idea of morphological wavelet more concrete, a classic morphological wavelet called the morphological Haar wavelet (MHW) was introduced by [60]. MHW is different from the classical linear Haar wavelet in that the linear signal analysis operator of the classical Haar wavelet is replaced by a morphological operator, e.g. taking the minimum or maximum over two samples.

Let $V_j = V_{j+1} = W_{j+1} = \Re^{\mathbb{Z}}$ ($j \geq 0$) be the lattice of doubly infinite real-valued sequences. In [60], the analysis and synthesis operators are defined as:

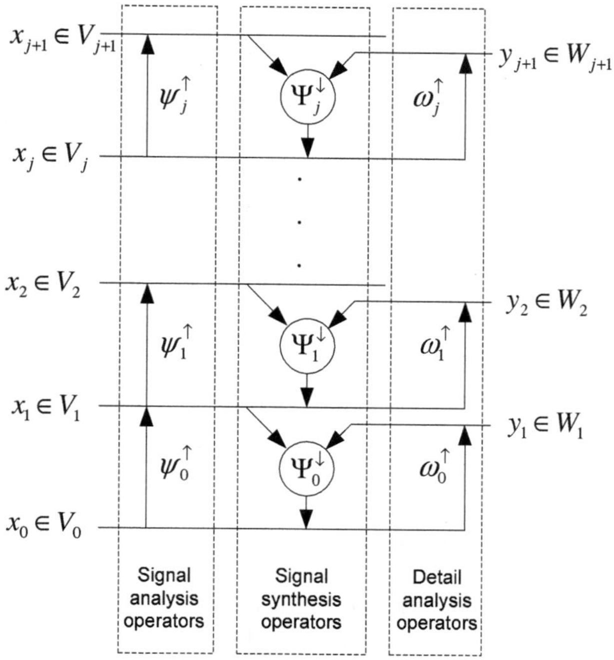

Fig. 2.8 Sample stages of the morphological wavelet decomposition scheme

$$\psi^{\uparrow}(x)(n) = x(2n) \wedge x(2n+1),\ x \in V_j, \tag{2.55}$$
$$\omega^{\uparrow}(x)(n) = x(2n) - x(2n+1),\ x \in V_j, \tag{2.56}$$

$$\begin{aligned} \psi^{\downarrow}(x)(2n) &= x(n),\ x \in V_{j+1}, \\ \psi^{\downarrow}(x)(2n+1) &= x(n),\ x \in V_{j+1}, \end{aligned} \tag{2.57}$$

$$\begin{aligned} \omega^{\downarrow}(y)(2n) &= y(n) \vee 0,\ y \in W_{j+1}, \\ \omega^{\downarrow}(y)(2n+1) &= -(y(n) \wedge 0),\ y \in W_{j+1}, \end{aligned} \tag{2.58}$$

$$\Psi^{\downarrow}(x,y)(n) = \psi^{\downarrow}(x)(n) + \omega^{\downarrow}(y)(n),\ x \in V_{j+1}, y \in W_{j+1}, \tag{2.59}$$

where n is the sample index, '$\wedge$' denotes the minimisation operation and '$\vee$' denotes the maximisation operation. Pseudo codes of the decomposition and reconstruction processes of MHW are listed in Tables 2.6 and 2.7, respectively. It is assumed here that the input signal is decomposed up to level K.

Table 2.6 The pseudo code of the decomposition process of MHW

```
Denote the input signal by x_0(n). Set k := 0;
while (k ≤ K)
  Get the analysis signal at level k+1:  x_{k+1}(n) = x_k(2n) ∧ x_k(2n+1);
  Get the synthesis signal at level k+1: y_{k+1}(n) = x_k(2n) − x_k(2n+1);
  k := k+1;
end
```

Table 2.7 The pseudo code of the reconstruction process of MHW

```
Set k := K;
while (k > 0)
  Get the even samples at level k−1: x_{k−1}(2n) = x_k(n) + (y_k(n) ∨ 0);
  Get the odd samples at level k−1:  x_{k−1}(2n+1) = x_k(n) − (y_k(n) ∧ 0);
  k := k−1;
end
```

2.4.3.2 The Morphological Gradient Wavelet

The morphological gradient wavelet (MGW) is a morphological wavelet that aims to take into consideration the gradient of each sample of the signal. Derived from [48, 66], a morphological median operator termed by $\overset{\mathrm{m}}{\oplus}$ is defined as:

$$(f \overset{\mathrm{m}}{\oplus} g)(x) = \mathrm{median}\{f(x-s)+g(s) | (x-s) \in \mathscr{D}_f, s \in \mathscr{D}_g\}, \tag{2.60}$$

where f is the signal, g is the SE, and $\mathscr{D}_f$ and $\mathscr{D}_g$ represent the definition domains of f and g, respectively. Apparently, instead of applying local minimum and maximum operations as dilation and erosion do, the morphological median operator returns the median value.

If the morphological median operator employs a flat SE whose length is 3 and whose origin is in the middle, (2.60) becomes:

$$(f \overset{\mathrm{m}}{\oplus} g)(x) = \mathrm{median}\{f(x-1), f(x), f(x+1)\}. \tag{2.61}$$

In MGW, the above equation is selected as the signal analysis operator, which is rewritten in (2.62). Consequently, the detail analysis operator of MGW is defined as depicted in (2.63).

$$\begin{aligned} \psi^{\uparrow}(x_j)(n) &= \mathrm{median}\{x_j(2n-1), x_j(2n), x_j(2n+1)\} \\ &= x_{j+1}(n), \end{aligned} \tag{2.62}$$

$$\begin{aligned} \omega^{\uparrow}(x_j)(n) &= x_j(2n+1) - 2x_{j+1}(n) + x_j(2n-1) \\ &= y_{j+1}(n). \end{aligned} \tag{2.63}$$

In order to construct the synthesis operator, two intermediate variables are needed as follows:

$$t_{\mathrm{R},j+1}(n) = x_j(2n+1) - x_j(2n), \tag{2.64}$$

$$t_{\mathrm{L},j+1}(n) = x_j(2n) - x_j(2n-1), \tag{2.65}$$

where $t_{\mathrm{R},j+1}(n)$ and $t_{\mathrm{L},j+1}(n)$ denote the right and left gradient of $x_j(2n)$, respectively. The synthesis operator is shown in (2.66)–(2.68). Pseudo codes of the decomposition and reconstruction processes of MGW are listed in Tables 2.8 and 2.9, respectively. It is also assumed that the input signal is decomposed to level K.

$$\left.\begin{aligned} \Psi_j^{\downarrow}(2n) &= x_{j+1}(n) \\ \Psi_j^{\downarrow}(2n+1) &= x_{j+1}(n) + \tfrac{1}{2}(y_{j+1}(n) + \\ &\quad t_{\mathrm{R},j+1}(n) + t_{\mathrm{L},j+1}(n)) \end{aligned}\right\} \text{ if } t_{\mathrm{R},j+1} \cdot t_{\mathrm{L},j+1} \geq 0; \tag{2.66}$$

$$\left.\begin{aligned} \Psi_j^{\downarrow}(2n) &= x_{j+1}(n) + t_{\mathrm{L},j+1}(n) \\ \Psi_j^{\downarrow}(2n+1) &= x_{j+1}(n) + y_{j+1}(n) \end{aligned}\right\} \text{ if } \begin{matrix} t_{\mathrm{R},j+1} \cdot t_{\mathrm{L},j+1} < 0 \text{ and} \\ |t_{\mathrm{R,j}+1}(n)| \geq |t_{\mathrm{L},j+1}(n)| \end{matrix}; \tag{2.67}$$

$$\left.\begin{aligned} \Psi_j^{\downarrow}(2n) &= x_{j+1}(n) + y_{j+1}(n) \\ &\quad + t_{\mathrm{L},j+1}(n) \\ \Psi_j^{\downarrow}(2n+1) &= x_{j+1}(n) \end{aligned}\right\} \text{ otherwise.} \tag{2.68}$$

Table 2.8 The pseudo code of the decomposition process of MGW

Denote the input signal by $x_0(n)$. Set $k := 0$;
while $(k \leq K)$
  Get the analysis signal at level $k+1$: $x_{k+1}(n) = \mathrm{median}\{x_k(2n-1), x_k(2n), x_k(2n+1)\}$;
  Get the synthesis signal at level $k+1$: $y_{k+1}(n) = x_k(2n+1) - 2x_{k+1}(n) + x_k(2n-1)$;
  Get the intermediate variables at level $k+1$: $t_{\mathrm{R},k+1}(n) = x_k(2n+1) - x_k(2n)$, $t_{\mathrm{L},k+1}(n) = x_k(2n) - x_k(2n-1)$;
  $k := k+1$;
end

Figure 2.9 shows the performance of MHW and MGW on a test signal [60]. To avoid misunderstanding, the approximation analysis operator and the detail analysis operator of MHW are denoted by $\psi_{\mathrm{H}}^{\uparrow}$ and $\omega_{\mathrm{H}}^{\uparrow}$, respectively, while those of MGW are denoted by $\psi_{\mathrm{G}}^{\uparrow}$ and $\omega_{\mathrm{G}}^{\uparrow}$, respectively. It can be seen that the detail signal obtained by $\omega_{\mathrm{H}}^{\uparrow}$ is zero at points where the input signal is constant, while that obtained by $\omega_{\mathrm{G}}^{\uparrow}$ is zero at points where the input signal is linear. Moreover, MGW not only detects every point where a change of gradient occurs, but also indicates the change quantitatively. These features make MGW suitable for the detection of sudden changes in

Table 2.9 The pseudo code of the reconstruction process of MGW

Set $k := K$;
while $(k > 0)$
  if $t_{R,k} \cdot t_{L,k} \geq 0$
    $x_{k-1}(2n) = x_k(n)$
    $x_{k-1}(2n+1) = x_k(n) + \frac{1}{2}(y_k(n) + t_{R,k}(n) + t_{L,k}(n))$
  else
    if $|t_{R,k}(n)| \geq |t_{L,k}(n)|$
      $x_{k-1}(2n) = x_k(n) + t_{L,k}(n)$
      $x_{k-1}(2n+1) = x_k(n) + y_k(n)$
    else
      $x_{k-1}(2n) = x_k(n) + y_k(n) + t_{L,k}(n)$
      $x_{k-1}(2n+1) = x_k(n)$
    end
  end
end

gradient. As it can be seen from Fig. 2.9e, the impulses explicitly reveal the location where the gradient changes. On the contrary, for MHW, changes of gradient in the original signal correspond to changes in the amplitude in the detail signal, as shown in Fig. 2.9c. The gradient information is not distinctly presented in the detail signal, which means that in this situation MGW is more applicable than MHW.

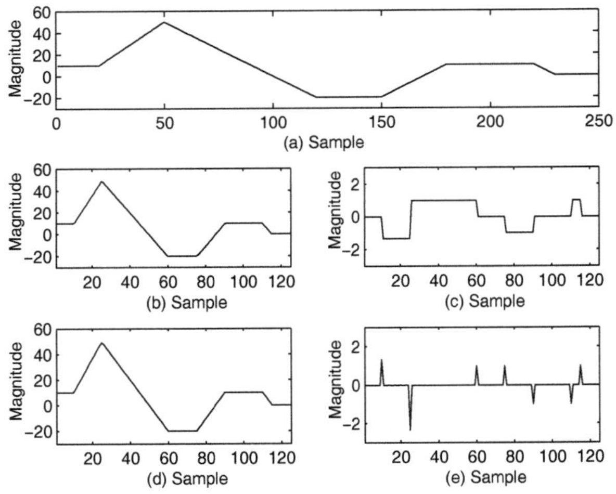

Fig. 2.9 The performance of MHW and MGW with analysis operators $\psi_H^{\uparrow}, \omega_H^{\uparrow}$ and $\psi_G^{\uparrow}, \omega_G^{\uparrow}$. **a** Original signal. **b** Approximation signal obtained by $\psi_H^{\uparrow}$. **c** Detail signal obtained by $\omega_H^{\uparrow}$. **d** Approximation signal obtained by $\psi_G^{\uparrow}$. **e** Detail signal obtained by $\omega_G^{\uparrow}$

2.4.4 The Multi-resolution Morphological Gradient Algorithm

The morphological gradient is defined as the arithmetic difference between the dilation and erosion of a signal by an SE. When using a flat SE, the morphological gradient is a powerful tool for edge detection. The multi-resolution morphological gradient (MMG) is an improved morphological gradient, which aims at depressing the steady state components and enhancing the transients ones. The multi-resolution morphological gradient algorithm (MMGA) introduced in this section is an algorithm designed to obtain the MMG of a signal, so that the ascending and descending edges of the transient waves can be extracted.

The morphological gradient is frequently used for edge detection in image and signal processing. Before giving its definition, two residual operators should be defined first. Denote the input signal and the SE by f and g, respectively. The dilation residual, G_{d}, is defined as the difference between the dilation of f by g and f, which is:

$$G_{\mathrm{d}}(f) = (f \oplus g) - f. \tag{2.69}$$

Similarly, we can define the erosion residual G_{e} to be the difference between f and the erosion of f by g, as follows:

$$G_{\mathrm{e}}(f) = f - (f \ominus g). \tag{2.70}$$

The morphological gradient is usually defined as:

$$G(f) = (f \oplus g) - (f \ominus g) = G_{\mathrm{d}}(f) + G_{\mathrm{e}}(f). \tag{2.71}$$

It should be noticed that there is a distinct meaning of morphological gradient, which is different from the gradient in physics. The geometric illustration and effect of the morphological gradient are depicted in Fig. 2.10, which demonstrates the process of the morphological gradient computation for a ramped-step signal, $f(x)$, by a flat SE, g, where the symbol Δ indicates the origin of the SE. Since dilation and erosion using a flat SE return the maximum and minimum of the neighbourhood, for every point, its morphological gradient reflects the difference between maximum and minimum obtained within the domain of the flat SE. Obviously, the morphological gradient is affected by the size and origin of the SE.

MMGA is designed based on the definition of the morphological gradient. In order to obtain the MMG, MMGA utilises a series of scalable flat SEs with different origins. The SEs are defined as:

$$g^{+} = \{0_1, 0_2, \ldots, 0_{l-1}, \underline{0_l}\}, \tag{2.72}$$

$$g^{-} = \{\underline{0_1}, 0_2, \ldots, 0_{l-1}, 0_l\}, \tag{2.73}$$

where g^{+} is the SE used for extracting the ascending edges and g^{-} is used for the descending edges of the signal; $l = 2^{1-a} l_g$, where a indicates the level of the MMG to be processed and l_g is the primary length of g at level 1; the underlined samples in g^{+} and g^{-} show their origins. Taking the definition in (2.71) as a reference, the

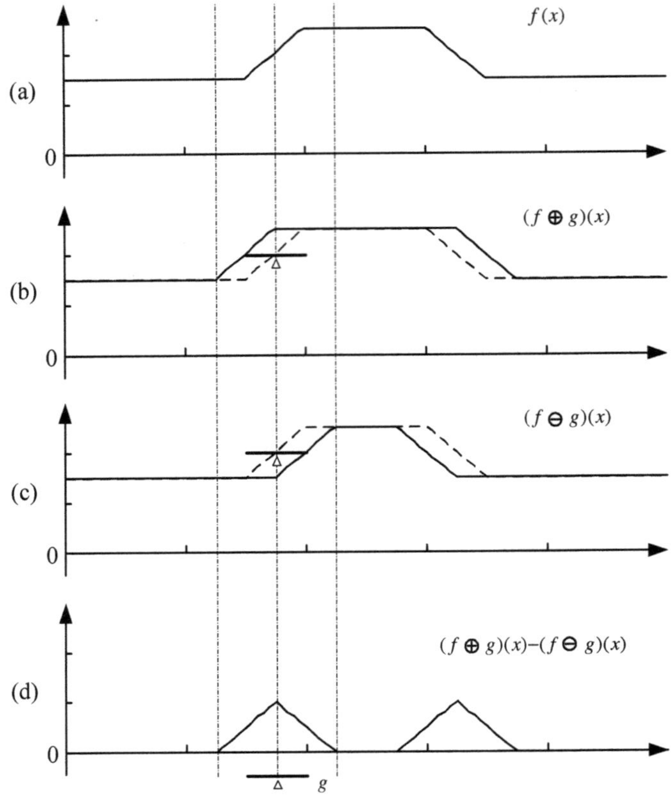

Fig. 2.10 The calculation and effect of the morphological gradient. **a** The input signal. **b** The dilation of f. **c** The erosion of f. **d** The morphological gradient

dyadic MMG at level a, ρ_g^a, is defined as:

$$\rho_{g^+}^a = (\rho^{a-1} \oplus g^+)(x) - (\rho^{a-1} \ominus g^+)(x), \tag{2.74}$$

$$\rho_{g^-}^a = (\rho^{a-1} \ominus g^-)(x) - (\rho^{a-1} \oplus g^-)(x), \tag{2.75}$$

$$\rho_g^a = \rho_{g^+}^a + \rho_{g^-}^a. \tag{2.76}$$

When $a = 1$, ρ^0 is the input signal. The pseudo code of MMGA is summarised in Table 2.10.

The explanations of (2.74) at level 1 is geometrically illustrated in Fig. 2.11. The solid lines in Figs. 2.11b and c are the results of dilation and erosion, respectively. For the purpose of comparison, the input signal is plotted with dashed lines. It is clear that under both transforms, the flat SE with the origin at its rightmost only affects the ascending edge of the input signal. This is because the origin of the flat SE should fall in the umbra of the input signal after either operation of dilation and erosion. Due to the geometrical symmetry of (2.74), there is no position shift-

Table 2.10 The pseudo code of MMGA

Denote the input signal by ρ^0. Determine the primary length l_g of the flat SE g. Set $a := 1$;
while (the end condition is not met)
  Construct the SEs:
    The length of the SE is $l_a = 2^{a-1} l_g$;
    $g^+(s) = \{0, 0, \cdots, 0\}$ and $1 - l_a \leq s \leq 0$;
    $g^-(s) = \{0, 0, \cdots, 0\}$ and $0 \leq s \leq l_a - 1$;
  Calculate the dilation and erosion of ρ^{a-1} by g^+ and g^-, respectively:
    $\delta_{a,+} = \rho^{a-1} \oplus g^+$
    $\delta_{a,-} = \rho^{a-1} \oplus g^-$
    $\varepsilon_{a,+} = \rho^{a-1} \ominus g^+$
    $\varepsilon_{a,-} = \rho^{a-1} \ominus g^-$
  Calculate the gradient:
    $\rho^a_{g+} = \delta_{a,+} - \varepsilon_{a,+}$
    $\rho^a_{g-} = \varepsilon_{a,-} - \delta_{a,-}$
    $\rho^a_g = \rho^a_{g+} + \rho^a_{g-}$
  $a := a + 1$;
end

ing introduced in the result illustrated in Fig. 2.11d. Likewise, the process of using MMGA to extract descending edges from the input signal is demonstrated in Fig. 2.12. In this case, the origin of the SE is at its leftmost as defined in (2.75) and MMGA is performed at level 1.

According to the ordering relation $(\rho^{a-1} \oplus g)(x) \geq (\rho^{a-1} \ominus g)(x)$, we have $\rho^a_{g+} \geq 0$ in (2.74) and $\rho^a_{g-} \leq 0$ in (2.75), which correspond to the ascending and descending edges of the signal waveform. Thus, ρ^a_g in (2.76) is able to provide the information of not only the exact location of waveform changes but also their polarities (changing directions). An example to illustrate this is shown in Fig. 2.13.

2.5 Summary

This chapter has introduced in detail the fundamental concepts of MM and its operations and features. The basic morphological operators of dilation and erosion have been defined, both in binary and in grey scale. Some of their properties have also been presented, which may help readers understand the operators and further deliver the ideas of morphological filtering, opening and closing. The mathematical calculations involved in morphological operations include only addition, subtraction, maximum and minimum operations without any multiplication and division. Besides, MM uses a much smaller size of the sampling window in real-time signal processing, which is different from integral transform-based algorithms that require a period of the signal to obtain its main features. Since it is a time-domain signal processing method and does not perform any integral transforms, MM is applicable to non-periodic transient signals and not restricted to periodic signals. A more ex-

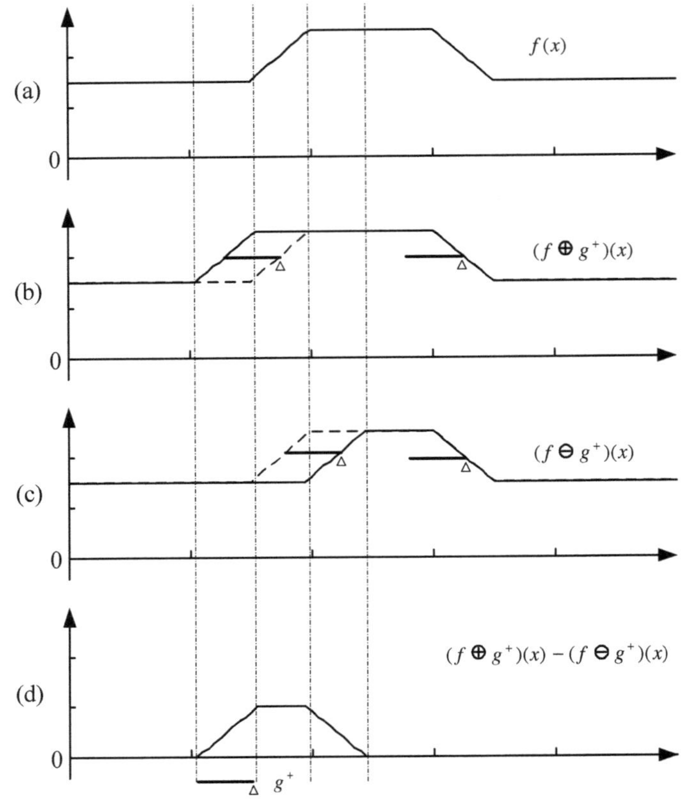

Fig. 2.11 The morphological gradient obtained by a flat SE with the origin at its rightmost. **a** The input signal. **b** The dilation of f. **c** The erosion of f. **d** The gradient at the first level

tensive discussion on MM can be found in the books by Matheron [99], Serra [125] and Soille [131], respectively.

MM has been mostly used in image processing; however, this book investigates MM for signal processing in particular. A large part of this chapter has been devoted to multi-resolution morphological operators, which are based on the framework of the multi-resolution decomposition scheme. This scheme provides us with an opportunity to view a signal at different levels. This chapter has introduced some classic operators, such as the lifting scheme, the morphological wavelet and MHW, as well as some operators that are derived from the classic methods, including MGW and MMGA. When acting upon signals of complex shapes, multi-resolution morphological operators are capable of decomposing a signal into meaningful parts and separating them from the background, while preserving the main shape characteristics.

In the following chapters, this book will show the application of MM to the development of novel algorithms for a variety of power system apparatus, including buses, transmission lines, and current transformers, etc. We will see that a group of

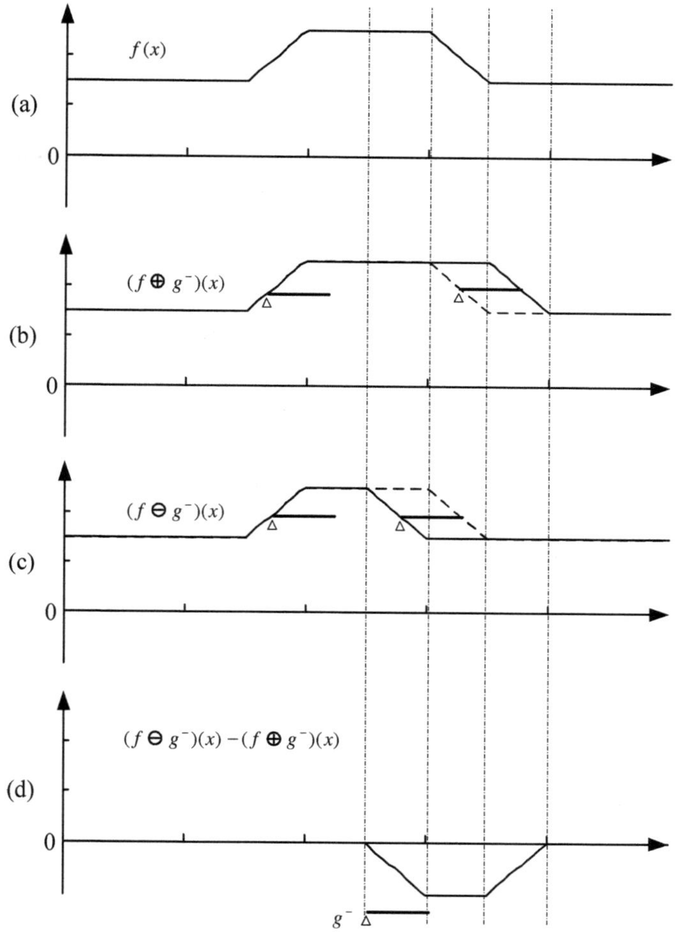

Fig. 2.12 The morphological gradient obtained by a flat SE with the origin at its leftmost. **a** The input signal. **b** The dilation of f. **c** The erosion of f. **d** The gradient at the first level

morphological operators have been developed specifically for protection algorithms, based on the operators and schemes introduced in this chapter.

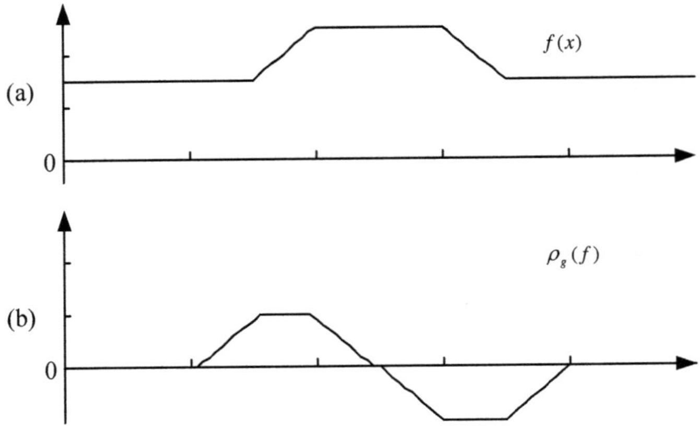

Fig. 2.13 A signal and its MMG at level 1. **a** The input signal. **b** Its MMG at level 1

Chapter 3
Phasor Measurement

Abstract The Fourier transform (FT) is commonly used for signal processing and in particular is adopted by digital distance protection relays for phasor measurement of electric power signals [139]. However, when a fault occurs on transmission lines, the input current of a relay may contain exponentially decaying DC offsets, which cannot be effectively removed by an FT-based filter. Therefore, fairly large errors are brought to the results of phasor measurement, which will cause malfunctioning of relays. In this chapter, a morphological transform (MT) is introduced to pre-process fault signals and remove the exponentially decaying DC offsets before they are processed by FT to perform phasor measurement [90].

3.1 Introduction

Fault signals taken by a relay contain large harmonics and exponentially decaying DC offsets which ought to be excluded and only the quantities of the fundamental frequency components of these signals should be retained for phasor measurement. FT is one of the most popular algorithms used for a variety of measurements in protection systems. It can be used to accurately extract the harmonic components from the signals if the following assumptions are satisfied:

- The highest frequency of the harmonic components of the input signal is less than $(N/2)f_1$, where N is the number of sampling points per fundamental cycle and f_1 is the fundamental frequency.
- There is no exponentially decaying DC offset.

Practically, a well-designed low-pass filter can be used to meet the requirement of the first assumption, but it is not likely to satisfy the second assumption under a fault condition. The exponentially decaying DC offsets contained in fault signals can introduce fairly large errors to phasor measurement and it is difficult to remove them accurately. With the existence of the exponentially decaying DC offset, the amplitude of the fundamental frequency and harmonic component, calculated by

FT, may deviate from the expected value by more than 15% in the worst case. Such a large error would highly reduce the accuracy of protection relays. Therefore, the exponentially decaying DC offsets need to be removed from the fault signals before they are processed by FT.

An algorithm using a full-cycle discrete Fourier transform (DFT) to calculate and compensate the exponentially decaying DC offset was presented in [56]. The algorithm requires a data window with a length of one fundamental cycle plus two samples to identify the DC offset and undertakes complex calculation, which obstructs its real-time application, In [119], the exponentially decaying DC offset was taken into account without knowing its time constant. The first two terms of the Taylor series expansion were used to represent the exponentially decaying DC offset, and the least squares error (LSE) curve fitting technique was applied to estimate the fundamental frequency component and other harmonics. Other approaches such as the Kalman filter were also evaluated in [12]. However, most of these approaches require a priori knowledge of the system parameters to estimate the exponentially decaying DC offset.

This chapter introduces the characteristics of fault voltages and currents, followed by presenting the algorithms that are conventionally adopted by protection relays, including FT. Aiming at reducing the FT error caused by the exponentially decaying DC offset, MT is presented to extract it from a fault signal. Thanks to this scheme, the accuracy of the calculation of the fundamental frequency component using DFT is greatly improved.

3.2 Phasor Measurement Methods

A protective relaying algorithm continuously processes samples sequentially collected from the measurement of instantaneous voltages and currents. The algorithm is required to extract amplitudes and phase angles from the sampled voltages and currents and compare them with the preset thresholds to identify the fault.

3.2.1 The Mann and Morrison Algorithm

The Mann and Morrison algorithm [98] only uses one sample and the derivative of the signal to calculate its amplitude and phase, based on the assumption that the voltage and current remain as a sinusoidal waveform after the fault occurs [117].

If the samples are obtained from a sinusoidal signal described by:

$$v(t) = V\sin(\omega_0 t), \tag{3.1}$$

where V is the amplitude of the voltage and $\omega_0 = 2\pi f_0$ and f_0 is the fundamental frequency. Then, the derivative of the signal is:

$$v'(t) = \omega_0 V \cos(\omega_0 t), \tag{3.2}$$

where $v'(t) = \mathrm{d}v(t)/\mathrm{d}t$. From the above equations, the peak value of the sinusoidal signal can be expressed as:

$$V = \sqrt{v(t)^2 + \left(\frac{v'(t)}{\omega_0}\right)^2}. \tag{3.3}$$

The above equations are assumed to be valid for any time instant. The derivative of the signal can be obtained from two consecutive samples if the signal is considered smooth and linear within a small interval. The derivative can be estimated as the slope of the linear segment of the signal and calculated by:

$$v'[k] = \frac{\Delta v}{\Delta t} = \frac{v[k+1] - v[k]}{\Delta t}, \tag{3.4}$$

where Δt is the sampling interval.

The amplitude and phase angle of the signal can therefore be calculated and updated by the following equations:

$$\begin{aligned} V[k] &= \sqrt{v[k]^2 + \left(\frac{v[k+1] - v[k]}{\omega_0 \Delta t}\right)^2}, \\ \phi[k] &= \tan^{-1}\left(\frac{v[k]\omega_0 \Delta t}{v[k+1] - v[k]}\right). \end{aligned} \tag{3.5}$$

A significant disadvantage of the Mann and Morrison algorithm is that the length of the sampling window is so short that the output of the algorithm would be unstable when the signal sampling suffers from any disturbances caused by deviation of sampling frequency or sampling instants.

3.2.2 The Rockefeller and Udren Algorithm

The Rockefeller and Udren algorithm uses both the first and second derivatives, which requires three samples, i.e. the sampling window is one sample longer than that used in the Mann and Morrison algorithm. Employing the same notations used in the Mann and Morrison algorithm, the first and second derivatives of the sinusoidal signal are expressed as:

$$v'[k] = \frac{v[k+1] - v[k-1]}{2\Delta t}, \tag{3.6}$$

$$v''[k] = \frac{v[k+1] - 2v[k] + v[k-1]}{(\Delta t)^2}. \tag{3.7}$$

The amplitude and phase angle of the sampled signal are:

$$V[k] = \frac{1}{\omega_0}\sqrt{(v'[k])^2 + \left(\frac{v''[k]}{\omega_0}\right)^2}, \tag{3.8}$$

and

$$\phi[k] = -\tan^{-1}\left(\frac{v''[k]}{\omega_0 v'[k]}\right). \tag{3.9}$$

The Rockefeller and Udren algorithm utilises only three samples, hence, it suffers from the same problem as the Mann and Morrison algorithm does. Both algorithms are extremely sensitive to a deviation of the sampling rate. It is impossible to use these algorithms to process non-sequentially sampled signals. In order to compensate the errors caused by non-sequentially sampled or delayed signals, algorithms that involve more samples are required. In fact, the most stable algorithm, which is widely used for protection relays, is DFT [38, 59].

3.2.3 The Full-Cycle Fourier Transform

The full-cycle FT is the most widely used algorithm for extraction of the amplitude and phase angle of a fault signal. It is immune to constant DC offsets and can filter integral harmonics [75, 147].

Any measured periodic voltage signal can be expanded into its Fourier series expansion [84] as:

$$v(t) = a_0 + \sum_{n=1}^{\infty} a_n \cos(n\omega_0 t) + \sum_{n=1}^{\infty} b_n \sin(n\omega_0 t), \tag{3.10}$$

where $\omega_0 = 2\pi f_0$ and f_0 is the fundamental frequency. The coefficients a_0, a_n and b_n can be obtained from:

$$a_0 = \frac{1}{T}\int_{t_0}^{t_0+T} v(t)\mathrm{d}t, \tag{3.11}$$

$$a_n = \frac{2}{T}\int_{t_0}^{t_0+T} v(t)\cos(n\omega_0 t)\mathrm{d}t, \quad n = 1, 2, \ldots, \infty, \tag{3.12}$$

$$b_n = \frac{2}{T}\int_{t_0}^{t_0+T} v(t)\sin(n\omega_0 t)\mathrm{d}t, \quad n = 1, 2, \ldots, \infty, \tag{3.13}$$

where T is the period of the fundamental frequency component of the signal. If the sampled signal is represented in a discrete form with N samples per fundamental cycle, the real and imaginary parts of the phasor representing the sampled signal are obtained as:

$$V_{\mathrm{Re}}(k) = \frac{2}{N}\sum_{n=0}^{N-1} v[k-n]\cos\left(\frac{2\pi n}{N}\right), \tag{3.14}$$

$$V_{\mathrm{Im}}(k) = \frac{2}{N}\sum_{n=0}^{N-1} v[k-n]\sin\left(\frac{2\pi n}{N}\right). \tag{3.15}$$

The amplitude and phase angle can be obtained using the real and imaginary components, respectively, as follows:

$$V(k) = \sqrt{(V_{\mathrm{Re}}(k))^2 + (V_{\mathrm{Im}}(k))^2}, \tag{3.16}$$

$$\phi[k] = \tan^{-1}\left(\frac{V_{\mathrm{Im}}(k)}{V_{\mathrm{Re}}(k)}\right). \tag{3.17}$$

3.2.4 The Half-Cycle Fourier Transform

To reduce the computation time of the algorithm by half, samples obtained from half a fundamental cycle are used. Hence, the coefficients of the Fourier series expansion are expressed by:

$$a_0 = \frac{1}{T/2}\int_{t_0}^{t_0+T/2} v(t)\mathrm{d}t, \tag{3.18}$$

$$a_n = \frac{2}{T/2}\int_{t_0}^{t_0+T/2} v(t)\cos(n\omega_0 t)\mathrm{d}t, \quad n = 1,2,\ldots,\infty, \tag{3.19}$$

$$b_n = \frac{2}{T/2}\int_{t_0}^{t_0+T/2} v(t)\sin(n\omega_0 t)\mathrm{d}t, \quad n = 1,2,\ldots,\infty, \tag{3.20}$$

where T is the period of the fundamental frequency component of the signal. Employing the same approach as used in the full-cycle FT, the real and imaginary parts of the phasor representing the sampled signal are calculated as:

$$V_{\mathrm{Re}}(k) = \frac{4}{N}\sum_{n=0}^{N/2-1} v[k-n]\cos\left(\frac{2\pi n}{N}\right), \tag{3.21}$$

$$V_{\mathrm{Im}}(k) = \frac{4}{N}\sum_{n=0}^{N/2-1} v[k-n]\sin\left(\frac{2\pi n}{N}\right). \tag{3.22}$$

Similarly, the amplitude and phase angle can be obtained by replacing the real and imaginary components in (3.16) and (3.17). The disadvantage of the half-cycle FT is that it is not able to filter out the exponentially decaying DC offset and even harmonics contained in the fault signal.

3.2.5 The Least Squares Error Algorithm

The LSE-based protective relaying algorithm was introduced by Sachdev and Baribeau in 1979 [119]. If a sampled signal contains only odd harmonics, and a low-pass filter is applied to eliminate the high-frequency components, the sampled signal, containing the exponentially decaying DC offset, fundamental frequency component and third harmonic, can be expressed as:

$$v(t) = K_0 e^{-\frac{t}{\tau}} + K_1 \sin(\omega_1 t + \theta_1) + K_3 \sin(\omega_3 t + \theta_3), \tag{3.23}$$

where τ is the time constant of the exponential component; K_0, K_1 and K_3 are constant parameters; $\omega_1 = 2\pi f_1$ and $\omega_3 = 2\pi f_3$ with f_1 the fundamental frequency and f_3 the frequency of the third harmonic; θ_1 and θ_3 are the initial phasor angle of the fundamental frequency component and the third harmonic, respectively. Using the first three elements of the Taylor series expansion, (3.23) can be rewritten as:

$$\begin{aligned} v(t) &= K_0 - K_0 t/\tau + K_0 t^2/2\tau^2 + K_1 \sin(\omega_1 t + \theta_1) + K_3 \sin(\omega_3 t + \theta_3) \\ &= rK_0 - K_0 t/\tau + K_0 t^2/2\tau^2 \\ &\quad + K_1 \sin(\omega_1 t)\cos\theta_1 + K_1 \cos(\omega_1 t)\sin\theta_1 \\ &\quad + K_3 \sin(\omega_3 t)\cos\theta_3 + K_3 \cos(\omega_3 t)\sin\theta_3 \end{aligned} . \tag{3.24}$$

With the following notations:

$$\begin{aligned} V &= [\, v_{k1} \;\; v_{k2} \;\; v_{k3} \;\; v_{k4} \;\; v_{k5} \;\; v_{k6} \;\; v_{k7} \,]^T \\ &= [\, 1 \;\; \sin(\omega_1 t_k) \;\; \cos(\omega_1 t_k) \;\; \sin(\omega_3 t_k) \;\; \cos(\omega_3 t_k) \;\; t_k \;\; t_k^2 \,]^T \end{aligned}$$

and

$$\begin{aligned} X &= [\, x_1 \;\; x_2 \;\; x_3 \;\; x_4 \;\; x_5 \;\; x_6 \;\; x_7 \,]^T \\ &= [\, K_0 \;\; K_1\cos\theta_1 \;\; K_1\sin\theta_1 \;\; K_3\cos\theta_3 \;\; K_3\sin\theta_3 \;\; -K_0/\tau \;\; K_0/2\tau^2 \,]^T, \end{aligned}$$

(3.24) can be written for its consecutive values at $t_1, t_2, \ldots, t_m$ as:

$$\begin{aligned} S &= [\, S_1 \;\; S_2 \;\; \cdots \;\; S_m \,]^T \\ &= \left[\sum_{n=1}^{7} v_{1n} x_n \;\; \sum_{n=1}^{7} v_{2n} x_n \;\; \cdots \;\; \sum_{n=1}^{7} v_{mn} x_n \right]^T . \end{aligned} \tag{3.25}$$

Then x_j can be obtained using the pseudo-inverse of matrix $V = [v_{ij}]$, where $i = 1, 2, \ldots, m$ and $j = 1, 2, \ldots, 7$. Equation (3.25) can be written in matrix form:

$$S = VX, \tag{3.26}$$

where $X = [x_j]$, $j = 1, 2, \ldots, 7$. Subsequently,

$$\begin{aligned} V^T S &= V^T V X, \\ (V^T V)^{-1} V^T S &= (V^T V)^{-1} V^T V X. \end{aligned}$$

As a result,

$$X = (V^T V)^{-1} V^T S. \tag{3.27}$$

Consequently, the amplitudes and phase angles of the fundamental frequency component and the third harmonic can be obtained from X.

3.3 Power System Faults

A typical power network is shown in Fig. 3.1, which illustrates the system with a fault occurring at point F. The characteristic of a fault current is related to the impedance of the transmission line and the amplitude and phase angle of the pre-fault current [9].

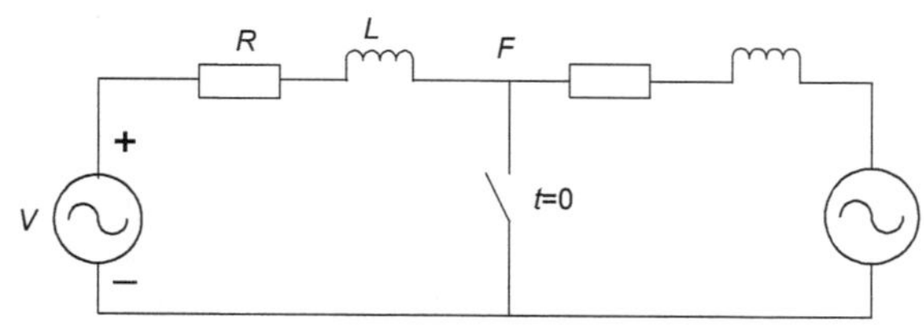

Fig. 3.1 A typical power network for fault simulation

3.3.1 Fault Characteristics

The transmission network is represented by resistance R and inductance L, which are Thevenin equivalent parameters [44] of the system, located to the left side of the fault point F as shown in Fig. 3.1. The differential equation of the circuit located on the left side is represented as:

$$L\frac{\mathrm{d}i}{\mathrm{d}t} + Ri = V_{\mathrm{m}} \sin(\omega t + \beta), \tag{3.28}$$

where V_{m} is the maximal value of the sinusoidal source voltage and β represents the angle of the supply voltage, at which the fault occurs. In the system, the shunt susceptances of the transmission line are neglected, which is an acceptable approximation in fault studies, since the voltages are severely depressed in the vicinity of the fault. However, this assumption needs to be reconsidered when applied to high-voltage systems.

The fault current can be simulated by solving the above equation and it is divided into two components, a steady-state component i_{s} and a transient component i_{t}:

$$\begin{aligned} i(t) &= i_s(t) + i_t(t) \\ &= \frac{V_m}{Z}\sin(\omega t + \beta - \theta) + \left[\frac{V_m}{Z}\sin(\theta - \beta) + i(0^+)\right] e^{-\frac{R}{L}t}. \end{aligned} \tag{3.29}$$

where $Z = \sqrt{R^2 + X^2}$, $X = \omega L$ and $\theta = \tan^{-1}\left(\frac{\omega L}{R}\right)$. The steady-state component, i_s, is in the same frequency as that of the source voltage, but it is shifted in phase by an angle of β and the constant angle of the system impedance, θ. The transient component, i_t, has two parts, one of which depends on angle β of the voltage waveform at which the fault occurs. The other part of the component is a function of the pre-fault current. In many cases, this pre-fault current is negligible compared with the transient fault current, which has a large amplitude.

3.3.2 *Influence of Exponentially Decaying DC Offset on FT Accuracy*

FT is frequently employed in protection relays to estimate the fundamental frequency component and to remove the exponentially decaying DC offset and the harmonics. However, a large exponentially decaying DC offset can make the amplitude of the fundamental frequency component estimated by FT to deviate from its real value. This error can be investigated from the power network shown in Fig. 3.1. In this network, the fault current, $i(t)$, can be represented as:

$$i(t) = A_0 e^{-\frac{t}{\tau}} + A_1 \sin(\omega t + \beta - \theta), \tag{3.30}$$

where $\tau = \frac{L}{R}$, $A_0 = \frac{V_m}{Z}\sin(\theta - \beta) + i(0^+)$ and $A_1 = \frac{V_m}{Z}$, and 0^+ denotes the time instant immediately after the fault occurs. Uniformly sampling the fault current every $\Delta T = 1/f$ under frequency f, a set of discrete values are obtained over a fundamental cycle, which are represented as:

$$i_k = A_0 r^k + A_1 \sin\left(\frac{2\pi}{N}k + \beta - \theta\right), \quad k = 0, \ldots, N, \tag{3.31}$$

where $r^k = e^{-\Delta T k/\tau}$, N is the number of sampling points per fundamental cycle. Applying FT to these discrete samples, the fundamental frequency components, $Z_{\text{real}}(k)$ and $Z_{\text{imag}}(k)$, can be obtained as follows:

$$Z_{\text{real}}(k) = \frac{2}{N} \sum_{r=k-N+1}^{k} i(r) \cos\left(\frac{2r\pi}{N}\right), \tag{3.32}$$

$$Z_{\text{imag}}(k) = \frac{2}{N} \sum_{r=k-N+1}^{k} i(r) \sin\left(\frac{2r\pi}{N}\right), \tag{3.33}$$

and the amplitude and phasor angle of the fundamental frequency component are calculated as:

$$A_k = \sqrt{Z_{\text{real}}(k)^2 + Z_{\text{imag}}(k)^2}, \tag{3.34}$$

$$\theta_k = \tan^{-1} \frac{Z_{\text{imag}}(k)}{Z_{\text{real}}(k)}. \tag{3.35}$$

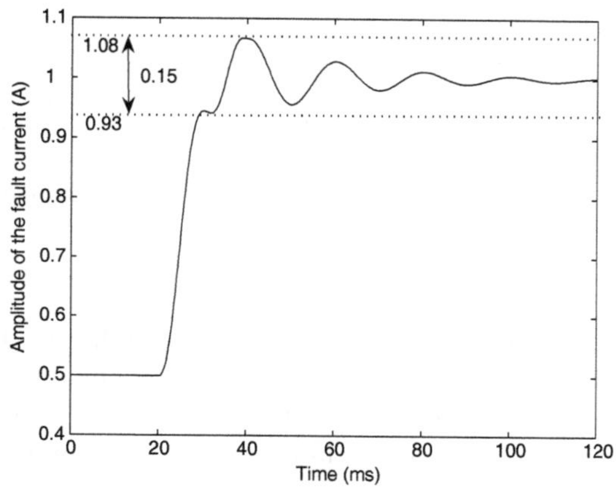

Fig. 3.2 The amplitude of the fundamental frequency component of the fault current, $i(t)$, estimated by FT

Figure 3.2 illustrates the amplitude of the fundamental frequency component of a fault current $i(t)$ that contains an exponentially decaying DC offset, which is calculated using FT. From Fig. 3.2, it can be observed that the maximal and minimal values calculated during the first cycle of the fault current may cause a relatively large error of more than 15%. Through the evaluation of the errors, a conclusion can be drawn that the presence of the exponentially decaying DC offset has a drastic impact on FT accuracy. Therefore, a filter should be designed to remove the exponentially decaying DC offset.

3.4 Morphological Transform for DC Offset Removal

MM is a non-linear approach for image and signal processing. In order to extract the exponentially decaying DC offset, the transient fault current is processed by the following MT:

$$y = (f \circ g \bullet g + f \bullet g \circ g)/2, \tag{3.36}$$

where f is the input current signal, g is the SE, y is the output of MT, and $\circ$ and $\bullet$ are the opening and closing operators defined in Sect. 2.3.2, respectively. This MT can

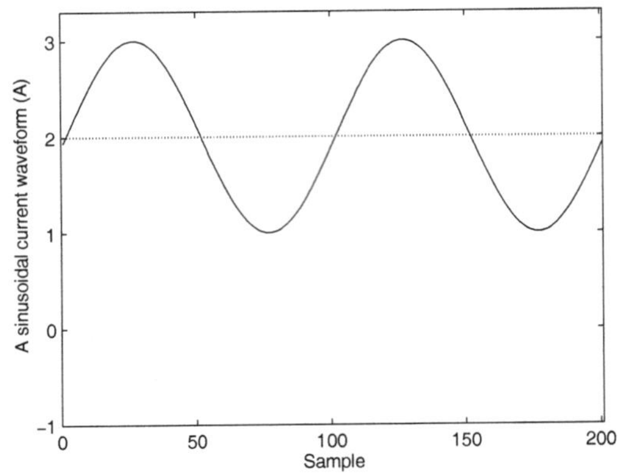

Fig. 3.3 *Solid line*: a sinusoidal waveform with a constant DC offset. *Dotted line*: the extracted DC offset by MT

remove all the detail waveforms whose widths are shorter than the length of the SE, if the SE is a flat one with its origin at the centre [134]. In this application, in order to extract the exponentially decaying DC offset, the waveforms of the fundamental frequency component and its higher harmonics should be removed from the input signal. Since the exponentially decaying DC offset varies much more slowly than the fundamental frequency component, and even much more slowly than the harmonics, the length of the SE should be half a fundamental cycle of the input signal. For example, for an input signal with the fundamental frequency of 50 Hz, the length of the SE should be 10 ms. For a signal with a 60 Hz fundamental frequency, it should be 1/120 s. In this manner, the fundamental frequency component and the harmonics are all removed from the input signal, and the output, y, is the DC offset. Figure 3.3 illustrates the extraction of a constant DC offset from the input signal.

Furthermore, in order to reduce calculation time, the samples obtained within a quarter-cycle of the waveform are used to extract the DC offset located within half a cycle, since the other quarter-cycle of the waveform can be symmetrically simulated. Therefore, an SE that has a half-cycle length is applied to extract the DC offset in each quarter-cycle of the input signal. The detailed procedure is illustrated in Fig. 3.4. As can be seen, the first quarter-cycle signal and its symmetric image, which is projected from the first quarter-cycle signal and used as the next quarter-cycle signal, are combined together to construct a half-cycle signal. Afterwards, the MT with the SE of the length of half a cycle is applied to extract the constant DC offset. Note that although the extracted DC offset is half a cycle long, only the first part is valid. Similarly, for the second quarter-cycle signal, its image is used as the first quarter-cycle signal to artificially construct a half-cycle signal. A constant DC offset can also be extracted from this reconstructed half-cycle signal by MT.

The DC offset of the whole signal can be extracted by applying MT quarter-cycle by quarter-cycle. In other words, the method outputs a constant DC offset every quarter-cycle in this case. The DC offset contained in the fault current, $i(t)$,

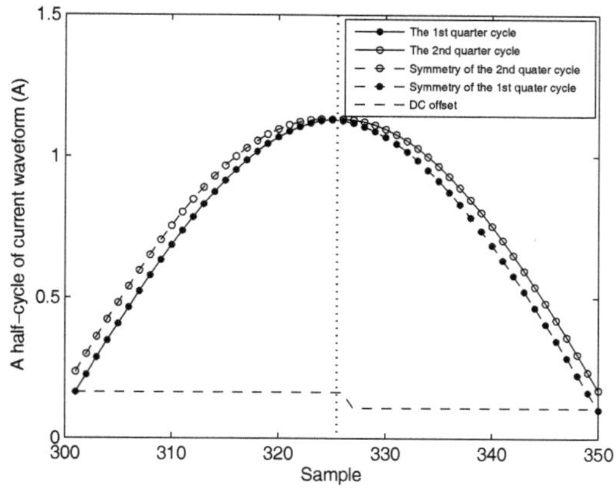

Fig. 3.4 Extraction of the DC offset from a half-cycle of the waveform and its symmetry

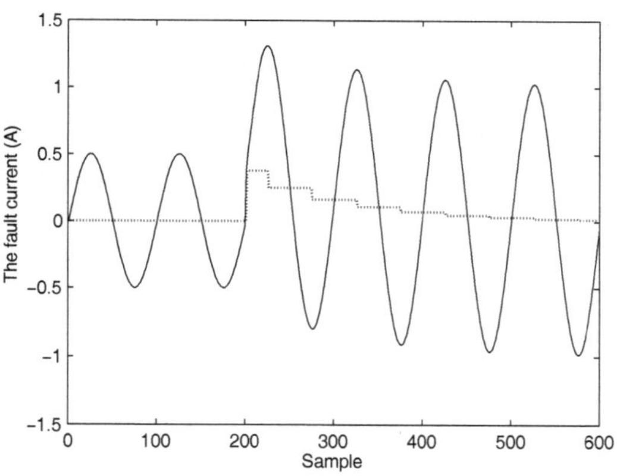

Fig. 3.5 *Solid line*: the input fault current. *Dotted line*: the DC offset extracted from fault current using MT

is presented by the dotted line in Fig. 3.5. It can be seen from the figure that there is difference between any two adjacent constant DC offsets. This indicates that an exponentially decaying DC offset exists in the input signal. Figure 3.6 illustrates a comparison of the exponentially decaying DC offset with the one extracted by MT. Apparently, the error between the exponentially decaying DC and the extracted DC offset is too large to be acceptable. Therefore, further actions are taken to rectify the extracted DC offset.

As the trajectory of an exponential function can be determined by three points, the first two points of the exponentially decaying DC offset obtained by MT can be used to estimate the third point, following the equation of an exponential function. Based on this equation, from the second half of the cycle, the exponentially decaying DC offset can be estimated at each sampling instant in a recursive manner, using the

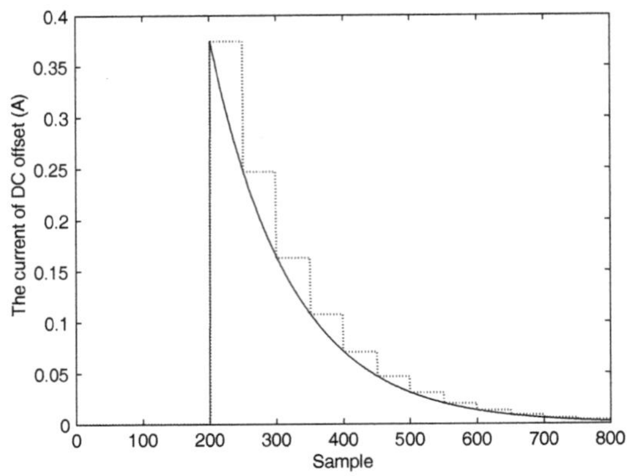

Fig. 3.6 *Solid line*: the exponentially decaying DC. *Dotted line*: the DC offset extracted by MT

information of the constant DC offsets extracted by MT from the previous half-cycle. The procedure is explained in detail as follows.

First, the first point of the exponentially decaying DC offset of the second half of the cycle is estimated using the two constant DC offsets obtained by performing MT in the first half of the cycle. A tilted line segment is constructed to approximate the exponentially decaying DC offset instead of the two horizontal line segments shown in Fig. 3.4. Its mathematical expression is determined by the first-order Taylor series expansion at $T_1/2$ over the interval $[0, T_1]$, which is represented as:

$$A_0 e^{-\frac{t}{\tau}} \cong C_0 + C_1 \left(t - \frac{T_1}{2} \right) = B_0 + B_1 t, \tag{3.37}$$

where T_1 represents the time interval of a fundamental cycle of the signal; C_0, C_1, B_0 and B_1 are coefficients. To determine B_0 and B_1, the values of the two horizontal line segments are needed. Denote them by K_1 and K_2 respectively and obtain:

$$\begin{aligned} K_1 &= B_0 + B_1 t, \\ K_2 &= B_0 + B_1 \left(t + \frac{N}{2} \Delta t \right), \end{aligned} \tag{3.38}$$

where N is the number of sampling points per fundamental cycle and $\Delta t = T_1/N$ is the time interval between two sampling points.

The derivatives of both sides of (3.37) are represented as:

$$\begin{aligned} &\frac{\mathrm{d}}{\mathrm{d}t}(A_0 e^{-\frac{t}{\tau}}) = -\frac{A_0}{\tau} e^{-\frac{t}{\tau}}, \\ &\frac{\mathrm{d}}{\mathrm{d}t}(B_0 + B_1 t) = B_1. \end{aligned} \tag{3.39}$$

Hereby, B_1 approximately equals the slope of the DC offset $-\frac{A_0}{\tau}e^{-\frac{t}{\tau}}$ when $t = 0$, i.e. $B_1 = -\frac{A_0}{\tau}e^{-\frac{0}{\tau}} = -\frac{A_0}{\tau}$.

Let B_2 denote the slope of the line segment to approximate the DC offset in the second half of the cycle. Its value amounts to the slope of the DC offset at $t = T_2/2$, which is:

$$B_2 = -\frac{A_0}{\tau}e^{-\frac{T_2/2}{\tau}} = B_1 e^{-\frac{T_2/2}{\tau}}. \tag{3.40}$$

Since $K_2 = K_1 e^{-\frac{T_2/2}{\tau}}$, B_2 can be calculated as $B_2 = K_2 B_1 / K_1$. Therefore, each point of the exponentially decaying DC offset in the second half of the cycle can be estimated by the values of B_2 and K_2.

This process is applied recursively to obtain the estimated exponentially decaying DC offset of the second half of the cycle. However, these estimates cannot be used to further estimate the exponentially decaying DC offset within the first half of the next cycle, because in this case the estimation error will be accumulated. Therefore, the samples of the second half of the cycle are collected to extract the constant DC offsets using MT, and the result will be used to correct the error of the estimated exponentially decaying DC offset at the beginning of the first half of the next cycle. The acquisition of a sample of the second half of the cycle takes place at the same time for MT to identify the DC offset when the corresponding point of the exponentially decaying DC offset is estimated.

After all the samples of the second half of the cycle have been collected, MT is applied to calculate its constant DC offsets and the result is then used to estimate the exponentially decaying DC offset of the first half of the next cycle. This process repeats, so that the exponentially decaying DC offset of the whole fault current period is estimated in real time, except for a half-cycle delay occurring at the very beginning.

However, the angle of the supply voltage does affect the delay of the protection action at the very beginning of fault occurrence. Here, the fault time instant is denoted as t_0 and the time instant at which the value of the fault signal reaches its maximal or minimal peak is denoted as t_p, which depends on the constant angle of the line impedance, θ, and the angle of the supply voltage at which the fault occurs, β. If $t_p - t_0 \leq 5$ ms, i.e. the duration is shorter than a quarter of a cycle, the protection action will not be triggered. Within this duration the fault signal and MT will not be processed, until the samples of the next quarter-cycle, counting from the first maximal or minimal peak, have been collected as a complete quarter-cycle waveform is required by MT to extract the DC offset. Therefore, the algorithm begins with a delay of up to a quarter of the cycle. Altogether there may be a delay of up to three quarters of a cycle from the very beginning of the fault before the exponentially decaying DC offset is estimated in each sampling interval. A flowchart is given in Fig. 3.7 to illustrate the procedure of the MT-based algorithm for the estimation of the exponentially decaying DC offset.

The estimation result of the exponentially decaying DC offset is shown in Fig. 3.8, in which the solid line is the estimated exponentially decaying DC offset. Look-

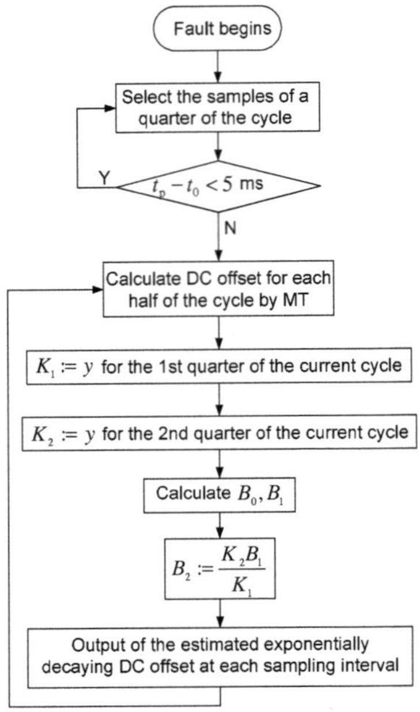

Fig. 3.7 Flowchart of the MT-based algorithm for estimation of the exponentially decaying DC offset

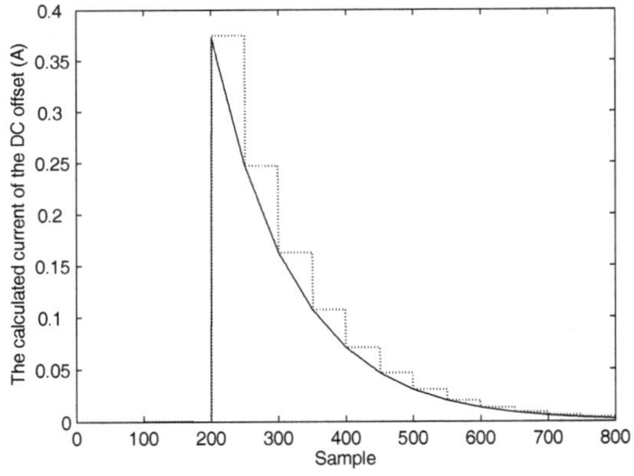

Fig. 3.8 *Dotted line*: the DC offset extracted by MT. *Solid line*: the estimated decaying DC offset that is calculated based on the signal plotted by the *dotted line*

ing closely at the solid line, one sees that it is composed of pieces of straight lines and is different from the solid line shown in Fig. 3.6, which is a smooth exponential signal. However, it can be seen that the difference between the simulated and estimated exponentially decaying DC offsets shown in Figs. 3.6 and 3.8 is negligible.

3.5 Results of Simulations and Discussions

Simulation studies are undertaken based on the power network illustrated in Fig. 3.1. A fault current is first processed by MT to remove the exponentially decaying DC offset, then the amplitude of the fundamental frequency component is calculated by FT. The result is presented by the solid line shown in Fig. 3.9. As a comparison, the result of applying FT directly to the input current is shown by the dotted line in the same figure.

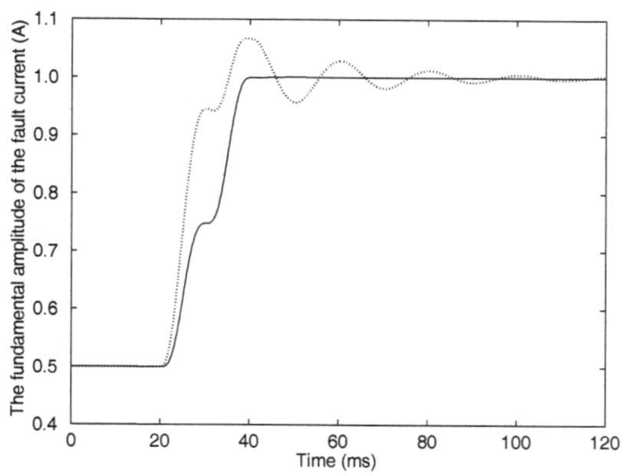

Fig. 3.9 The amplitude of the fundamental frequency component estimated by FT with (*solid line*) and without (*dotted line*) the removal of the exponentially decaying DC offset

Tests are carried out with different sampling rates. The results are listed in Table 3.1, where A_{max} and A_{min} denote the maximal and minimal amplitudes of the fundamental frequency components obtained using FT and MT plus FT, respectively.

From the simulation results shown in Table 3.1, it can be observed that the peak value only changes slightly for different sampling rates, which reflects that the sampling rate does not affect the accuracy of the estimation result. However, it is true that a lower sampling rate makes the estimated exponentially decaying DC offset less smooth. For each sampling rate, the MT-based method narrows the amplitude deviation. It reduces the overshoot of the maximal peak from 1.0664 to 1.0001; and for the minimum, it ascends from 0.9563 to 0.9997.

Other simulation studies are carried out by changing the angle of the supply voltage in Fig. 3.1, which is denoted as β in (3.28). Although the time instant, at which the peak value of the fault signal occurs, changes significantly with different β, the peak value barely changes. Similar results are obtained from the simulation studies undertaken in this case, which show that the estimation accuracy is not affected by angle of the supply voltage.

Table 3.1 The peak values at different sampling rates

Sampling rate	FT		MT+FT	
N	$A_{\max}$	$A_{\min}$	$A_{\max}$	$A_{\min}$
12	1.0612	0.9563	1.0031	0.9839
24	1.0650	0.9568	1.0011	0.9985
36	1.0659	0.9571	1.0005	0.9994
48	1.0661	0.9573	1.0004	0.9996
100	1.0664	0.9577	1.0001	0.9997

3.6 Summary

The current and voltage signals contain large harmonics and exponentially decaying DC offsets after a fault occurs. FT is normally used as a low-pass filter to filter high-order harmonics. However, it is unable to extract the exponentially decaying DC offsets of transient post-fault signals accurately using a short sampling window.

In this chapter, MT is applied to eliminate the influence of the exponentially decaying DC offset on FT. The MT first extracts the two constant DC offsets from the first and second quarter-cycle signals, respectively. Afterwards, the exponentially decaying DC offset of the second half of the cycle is estimated in each sampling interval, based on the constant DC offsets of the first half of the cycle. Meanwhile, the two constant DC offsets of the second half of the cycle are calculated, in order to estimate the exponentially decaying DC offset of the first half of the next cycle for the purpose of error correction. The above procedure repeats for the whole signal. This MT-based algorithm has a half-cycle delay in the best case and a $3/4$-cycle delay in the worst case at the very beginning after the fault occurrence. Afterwards, the exponentially decaying DC offset can be estimated in each sampling interval. As a matter of fact, the method can be used for all protective relaying algorithms requiring exponentially decaying DC offset removal. The evaluation of the algorithm's performance is carried out on fault current signals generated from the typical power network, which demonstrates that MT can extract the exponentially decaying DC offsets from the fault signal and improve the performance of FT used for phasor measurement, and it is not sensitive to different sampling rates.

Chapter 4
Protection of Transmission Lines

Abstract This chapter presents two examples of the MM-based method used for protection of transmission lines. One is concerned with an adaptive distance relaying algorithm (ADRA) designed for transmission line protection. In ADRA, a morphological fault detector (MFD) designed based on MM is used to determine the occurrence of a fault. The Euclidean norm of the output of the MFD is then calculated for fault phase selection and fault type classification. With respect to a specific type of fault scenario, an instantaneous circuit model applicable to a transient fault process is constructed to determine the fault location. The distance between the fault location and the relay position is calculated by a differential equation of the instantaneous circuit model, which is resolved in a recursive manner within each sampling interval. Due to the nature of recursive calculation, the protection zone of the relay varies from a small length to a large one, which increases as the accuracy of estimated fault distance improves when more samples are involved in the recursive calculation after the fault has been detected. The other example is concerned with the problem of mis-selection of a fault phase during a fault occurring on a double circuit transmission line. This problem is caused by the interaction between the phase lines of the two circuits, due to their mutual coupling. An MM-based fault phase selection scheme is introduced to accurately classify the fault types [92]. In this scheme, the fault characteristics can be effectively extracted from the associated currents by MGW. The fault phase selection scheme is evaluated on a simulation model under different fault types, locations, ground-fault resistances and inception angles.

4.1 Introduction

Distance protection relays have been widely applied for the protection of transmission lines for many years [109, 156]. The principle of distance protection is to measure the impedance, at a fundamental frequency, between the fault location and the protection relay, thus to determine whether a fault is internal or external to a protection zone. Conventionally, the impedance is calculated in real time to determine whether a fault location is within a protection zone. Meanwhile, a fault location algorithm is triggered after the relay action is taken for the purpose of fault reporting, using the pre-fault and post-fault data recorded. However, with the rapid development of microprocessor technology, it is possible to develop real-time fault location algorithms for distance protection. These algorithms use data gathered from the relay at one terminal of the transmission line and are called 'one terminal fault location algorithms' [65, 112].

Traditional impedance calculation for distance protection is based on the measurement of the fundamental phasors of voltage and/or current signals. When a fault occurs, the voltage and/or current signals are severely distorted and they may contain harmonics and exponentially decaying DC components. It is very difficult to quickly and accurately estimate the fundamental phasors from the measurement of these signals. Thus, the performance of distance protection relays may be deteriorated [109]. Moreover, the accuracy of distance protection relay is affected by a ground-fault resistance [39]. This resistance is not in a constant value in practice. Instead, it is related to various factors such as the voltage level, tower footing resistance, resistivity of soil and environmental conditions. The ground-fault resistance introduces an error in the impedance measurement and thus can affect the precision of distance protection relays.

The concept of adaptation has been introduced to protection relays [58, 121] and has received increasing interest during the last decade due to the advancement of computers, communication systems and software techniques [15]. The benefit brought by adaptive protective relaying arises from the integration with substation control and data acquisition functions and interfacing with the central energy management system [116]. Its advantages have been discussed in [25, 155].

A traditional distance protective relaying algorithm requires a data window with a length equivalent to one cycle of the fundamental frequency to calculate the fault impedance from current and voltage signals. It detects a fault approximately one cycle after the fault occurs. The size of short data windows used for impedance calculation would allow fast impedance estimation and fault detection [128]. However, the algorithms designed based on relatively short data windows generally do not meet the reliability requirement of relays used for modern transmission networks. In order to achieve the reliability as robust as that offered by the algorithms implemented using a full cycle of waveform, ADRA is introduced in this chapter to use a variable window length for impedance estimation. The occurrence of a fault is first detected by MFD, and then the fault is classified according to the components extracted by MFD. The estimation of the fault location is realised by a real-time fault location algorithm, which is developed based on an instantaneous circuit model. Moreover, the algorithm adopts a variable tripping zone, so that it is able to maximise the length of the first protection zone of the distance relay.

4.2 The Adaptive Distance Relaying Algorithm

4.2.1 Fault Detection

It is well known that a steady fault current can be expressed as a combination of two components [133]. One is a steady-state sinusoidal waveform determined by the source voltage and fault circuit impedance. The other is an exponentially decaying DC offset caused by the fault. The current signal, $i(t)$, has the form:

$$i(t) = A\cos(\omega t + \theta) + Be^{\lambda t}, \tag{4.1}$$

where ω and θ are the angular frequency and phase shift of the current, $i(t)$; A and B are amplitudes of the sinusoidal waveform and exponentially decaying DC offset, respectively. In (4.1), λ takes the value of $-1/\tau$, where τ is the time constant of the transmission line system where the fault occurs. The expression is based on the assumption that the current does not contain any harmonics. The current, $i(t)$, is sampled at a speed of N_s samples per fundamental cycle. The sampling interval is $\Delta t = T/N_s$, where T is the duration of a fundamental cycle. Then the kth sample of current $i(t)$ is represented as:

$$I(k) = A\cos(\omega \cdot k\Delta t + \theta) + Be^{\lambda k\Delta t}, \tag{4.2}$$

and its first-order approximation of the Taylor series is:

$$I(k) = A\cos(\omega \cdot k\Delta t + \theta) + B + \lambda k\Delta t. \tag{4.3}$$

Let I_k be the centre point at which the Taylor expansion takes place. The samples at its left and right are denoted as:

$$\begin{aligned} I(k+n) &= A\cos(\omega \cdot (k+n)\Delta t + \theta) + B + (k+n)\lambda\Delta t, \\ I(k-n) &= A\cos(\omega \cdot (k-n)\Delta t + \theta) + B + (k-n)\lambda\Delta t, \\ I(k+n) + I(k-n) &= 2I(k)\cos(\omega \cdot n\Delta t) + 2(B + \lambda k\Delta t) \cdot (1 - \cos(\omega \cdot n\Delta t)). \end{aligned}$$

If n is a small integer and the sampling frequency is high, the value of $\cos(\omega \cdot n\Delta t)$ approximates to 1, which leads to:

$$I(k+n) + I(k-n) \approx 2I(k)\cos(\omega \cdot n\Delta t). \tag{4.4}$$

In order to detect the occurrence of a fault on a transmission line, we introduce an intermediate operation, which involves morphological operation. The occurrence of a fault causes transient disturbances to the current or voltage waveform. The features of the disturbances can be captured by a morphological operator. In order to take into consideration the sinusoidal characteristics of current and voltage signals, an SE, b, is defined as follows:

$$b = [\cos m\phi,\ \ldots,\ \cos 2\phi,\ \cos\phi,\ 1,\ \cos\phi,\ \cos 2\phi,\ \ldots,\ \cos m\phi], \tag{4.5}$$

so that it can represent the sinusoidal feature. Here, $\phi = \omega\Delta t$, where ω is the fundamental angular frequency of the power system and Δt the sampling time interval. For a signal, $I(k)$, the dilation and erosion operators are defined as:

$$(I \oplus b)(k) = \max_{s}\{I(k-s)/b(s)\}, \tag{4.6}$$

$$(I \ominus b)(k) = \min_{s}\{I(k+s)/b(s)\}. \tag{4.7}$$

This type of dilation and erosion operators are called 'weighted dilation and erosion' [123]. In order to serve the purpose of fault detection, the operators are further developed in the following way. First, we define a group of SEs, $\{b_1,\ b_2,\ \ldots,\ b_n,\ \ldots,\ b_m\}$, as follows:

$$\begin{bmatrix} b_1 \\ b_2 \\ \vdots \\ b_m \end{bmatrix} = \begin{bmatrix} * & \cdots & * & \cos\phi & * & \cos\phi & * & \cdots & * \\ * & \cdots & \cos 2\phi & \cos\phi & * & \cos\phi & \cos 2\phi & \cdots & * \\ & & & & \vdots & & & & \\ \cos m\phi & \cdots & \cos 2\phi & \cos\phi & * & \cos\phi & \cos 2\phi & \cdots & \cos m\phi \end{bmatrix}, \tag{4.8}$$

where $*$ means that the corresponding sample of signal I is not involved in the morphological operation, m denotes the number of the SEs defined in the group, the nth SE has a length of $(2n+1)$, and $1 \le n \le m$. Accordingly, the dilation and erosion operators are defined as:

$$(I \oplus b_n)(k) = \max_{s}\{I(k-s)/b_n(s)\}, \tag{4.9}$$

$$(I \ominus b_n)(k) = \min_{s}\{I(k+s)/b_n(s)\}. \tag{4.10}$$

(4.11)

The intermediate operation, $D_n(k)$, is therefore defined as:

$$D_n(k) = \frac{1}{2}(I \oplus b_n + I \ominus b_n). \tag{4.12}$$

When $n = 1$, $D_1(k)$ is calculated as:

$$D_1(k) = \frac{I(k+1) + I(k-1)}{2\cos(\omega \cdot \Delta t)} \approx I(k). \tag{4.13}$$

The difference between $I(k)$ and D_n is estimated as:

$$\triangle I(k) = I(k) - \frac{D_1(k) + D_2(k)}{2}. \tag{4.14}$$

An MFD output, $\triangle D(k)$, is constructed based on $\triangle I(k)$, which is denoted as:

$$\triangle D(k) = |\triangle I(k+1) - \triangle I(k)|. \tag{4.15}$$

The value of $\triangle D(k)$ is monitored and compared with a preset threshold M. If $\triangle D(k) > M$, a counter C is initialised to 1. During the fault, as sampling proceeds step by step, counter C is increased by 1 if $\triangle D(k+1)$ is larger than M, while it is decreased by 1 if $\triangle D(k+1) \leq M$ and $C \geq 1$. The process can be represented as:

$$C := \begin{cases} C+1 : \triangle D(k+1) > M \\ C-1 : \triangle D(k+1) \leq M \text{ and C} \geq 1 \end{cases}. \tag{4.16}$$

The same process is simultaneously applied to all the phase currents on the transmission line. A fault occurrence is detected if counter C is greater than a threshold C_{set} for any phase current. In (4.16), M is preset with consideration of system operation conditions and C_{set} is given depending on the sampling frequency. The process of fault detection is described by the flowchart shown in Fig. 4.1.

4.2.2 Fault Classification

The zero-sequence current I_0 and the three phase currents $I_{\text{A}}, I_{\text{B}}$ and I_{C} are composed as the input signals for fault phase selection. In total, the four current signals are fed to (4.12) and (4.15), and their MFD outputs can be obtained, which are denoted as $\triangle D_i$, with $i = 0,1,2,3$, corresponding to I_0, I_{A}, I_{B} and I_{C}, respectively.

The Euclidean norm of $\triangle D_i$ is calculated to measure the amount of energy contained in $\triangle D_i$. The norm, denoted as E_i, can be calculated as follows:

$$E_i = \sqrt{\sum_{k=1}^{N_{\text{D}}} \triangle D_i^2(k)}, \quad i = 0,1,2,3, \tag{4.17}$$

where N_{D} is the number of samples contained in the window for the norm calculation. The discrimination components can be derived as follows:

$$L_i = \begin{cases} 1 & E_i > \varepsilon \\ 0 & E_i \leq \varepsilon \end{cases}, \tag{4.18}$$

where ε is a threshold value that can be determined considering the system operation condition, and L_i is a logic value that will be used as a discriminant component for fault phase selection. The phase selection is performed by measuring the norms of all the four input currents, since the norms of fault phases are much higher than those of healthy phases. If the norm exceeds a preset threshold value, its corresponding phase is detected as the fault phase. Figure 4.2 describes the process of fault classification.

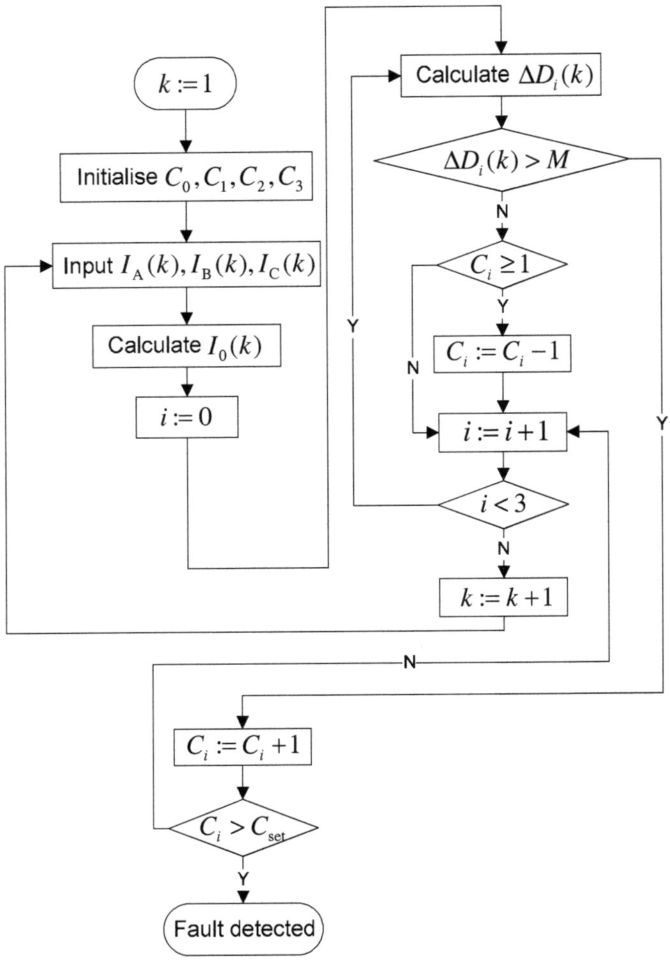

Fig. 4.1 A flowchart for fault detection

4.2.3 Fault Location

4.2.3.1 Single-Phase-to-Ground Fault

Suppose that a fault occurs on phase A with a ground-fault resistance of R_f on a transmission line at an unknown distance, D_f, measured from the relay position. The symmetrical components of the voltage of phase A at the relay position are represented as follows [108]:

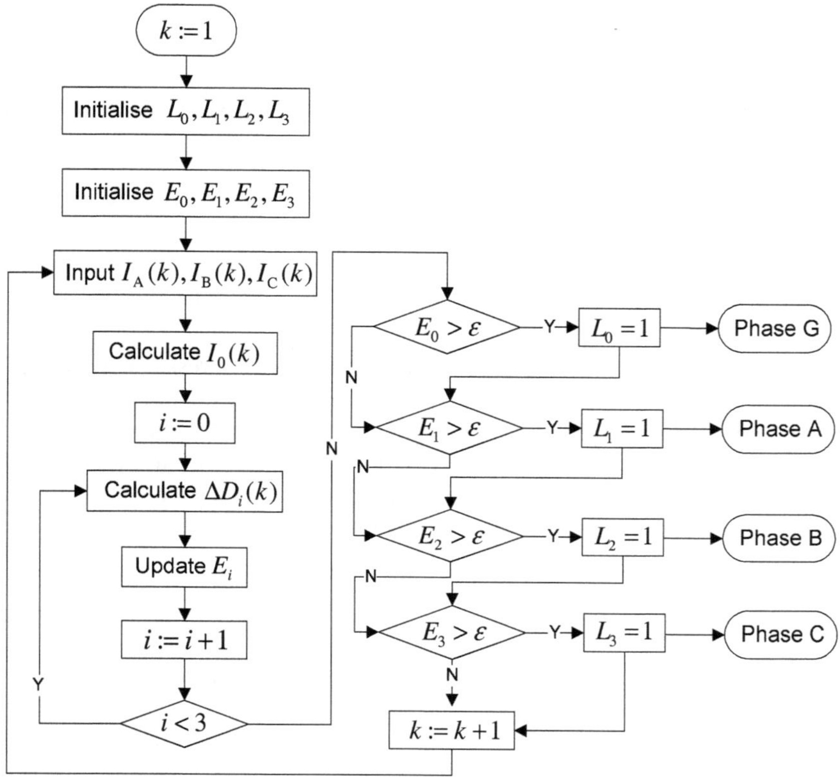

Fig. 4.2 A flowchart for fault classification

$$
\begin{aligned}
V_{a0} &= Z_0 I_{a0} + V^{\mathrm{f}}_{a0},\\
V_{a1} &= Z_1 I_{a1} + V^{\mathrm{f}}_{a1},\\
V_{a2} &= Z_2 I_{a2} + V^{\mathrm{f}}_{a2},
\end{aligned} \tag{4.19}
$$

where V_{a0}, V_{a1} and V_{a2} are the sequence components of phase A voltage at the relay position; I_{a0}, I_{a1} and I_{a2} are the sequence components of phase A current at the relay position; Z_0, Z_1 and Z_2 are the sequence impedances between the relay position and the fault location; and V^{f}_{a0}, V^{f}_{a1} and V^{f}_{a2} are the sequence components of phase A voltage at the fault location.

Summing the three equations above, we obtain:

$$V_a = (Z_0 - Z_1)I_0 + Z_1 I_a + V^{\mathrm{f}}_a, \tag{4.20}$$

where $Z_1 = Z_2$ has been assumed for the transmission line. Let $Z_1 = R_1 + sL_1$ and $Z_0 = R_0 + sL_0$, where s is a Laplace variable. In the s-domain, (4.20) becomes:

$$V_a(s) = (R_0 + sL_0 - R_1 - sL_1)I_0(s) + (R_1 + sL_1)I_a(s) + V_a^{\mathrm{f}}(s). \tag{4.21}$$

Applying the inverse Laplace transform to (4.21), the voltage of phase A at the relay position can be represented in the time domain as:

$$\begin{aligned} v_a(t) = v_a^{\mathrm{f}}(t) + D_{\mathrm{f}}\left[r_1 i_a(t) + l_1 \frac{\mathrm{d}i_a(t)}{\mathrm{d}t}\right] \\ + D_{\mathrm{f}}\left[(r_0 - r_1)i_0(t) + (l_0 - l_1)\frac{\mathrm{d}i_0(t)}{\mathrm{d}t}\right], \end{aligned} \tag{4.22}$$

where r_1 and r_0 are the positive-sequence and zero-sequence resistances in per unit length, respectively; l_1 and l_0 are the positive- and zero-sequence inductances in per unit length, respectively; $v_a(t)$ and $v_a^{\mathrm{f}}(t)$ are the phase A voltages at the relay position and fault location, respectively; $i_a(t)$ is the phase A current at the relay position, and $i_0^{\mathrm{f}}(t)$ is the zero-sequence current at the fault location.

The voltage at the fault location, $v_a^{\mathrm{f}}(t)$, can be expressed in terms of the ground-fault resistance R_{f} and the fault current $i_a^{\mathrm{f}}(t)$ as:

$$v_a^{\mathrm{f}}(t) = R_{\mathrm{f}} i_a^{\mathrm{f}}(t). \tag{4.23}$$

In the case of a single-phase-to-ground fault, the zero-sequence component of the fault current can be calculated by:

$$i_0^{\mathrm{f}}(t) = \frac{i_a^{\mathrm{f}}(t)}{3}. \tag{4.24}$$

Thus, $v_a^{\mathrm{f}}(t)$ can be denoted as:

$$v_a^{\mathrm{f}}(t) = 3R_{\mathrm{f}} i_0^{\mathrm{f}}(t). \tag{4.25}$$

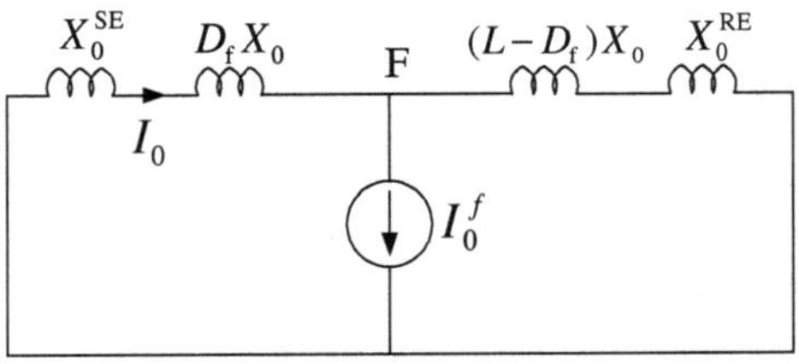

Fig. 4.3 The zero-sequence network for a single-phase-to-ground fault

According to the structure of the zero-sequence network as shown in Fig. 4.3, the zero-sequence fault current splits into two parts, which have the following relationship with the zero-sequence current at the relay position [7]:

$$i_0^{\mathrm{f}}(t) = i_0(t)\left[\frac{(x_0^{\mathrm{SE}} + D_{\mathrm{f}}x_0) + x_0^{\mathrm{RE}} + (L - D_{\mathrm{f}})x_0}{x_0^{\mathrm{RE}} + (L - D_{\mathrm{f}})x_0}\right] \\ = i_0(t)(1 + K_0), \tag{4.26}$$

where x_0^{SE} and x_0^{RE} are the zero-sequence impedances of the sending-end and remote-end sources, respectively; L is the length of the transmission line, and

$$K_0 = \frac{x_0^{\mathrm{SE}} + D_{\mathrm{f}}x_0}{x_0^{\mathrm{RE}} + (L - D_{\mathrm{f}})x_0}. \tag{4.27}$$

Thus, (4.25) can be rewritten as:

$$v_a^{\mathrm{f}}(t) = 3R_{\mathrm{f}}i_0(t)(1 + K_0). \tag{4.28}$$

Let

$$R_{\mathrm{f}}' = 3R_{\mathrm{f}}(1 + K_0), \tag{4.29}$$

then

$$v_a^{\mathrm{f}}(t) = R_{\mathrm{f}}' i_0(t). \tag{4.30}$$

Substituting (4.30) into (4.22) yields the required differential equation that models the instantaneous transmission line circuit during a phase-to-ground fault, which is given by:

$$v_a(t) = R_{\mathrm{f}}' i_0(t) + D_{\mathrm{f}}\left[r_1 i_a(t) + l_1 \frac{\mathrm{d}i_a(t)}{\mathrm{d}t}\right] \\ + D_{\mathrm{f}}\left[(r_0 - r_1)i_0(t) + (l_0 - l_1)\frac{\mathrm{d}i_0(t)}{\mathrm{d}t}\right]. \tag{4.31}$$

It can be written in the form of

$$v_a(t) = \left[H_{1a}(t)\ H_{2a}(t)\right]\begin{bmatrix} R_{\mathrm{f}}' \\ D_{\mathrm{f}} \end{bmatrix}, \tag{4.32}$$

where

$$H_{1a}(t) = i_0(t),$$

and

$$H_{2a}(t) = r_1 i_a(t) + (r_0 - r_1)i_0(t) \\ + \left[l_1 \frac{\mathrm{d}i_a(t)}{\mathrm{d}t} + (l_0 - l_1)\frac{\mathrm{d}i_0(t)}{\mathrm{d}t}\right].$$

For phase-B- or phase-C-ground faults, similar equations can be easily derived based on the above discussions.

4.2.3.2 Phase-to-Phase Fault

The ground-fault resistance is negligible for phase-to-phase faults and there is no zero-sequence current component. These types of faults can be described by a differential equation based on an instantaneous circuit model. For a phase-A-B fault, the differential equation is given as:

$$v_{ab}(t) = D_{\mathrm{f}} r_1 (i_a(t) - i_b(t)) + D_{\mathrm{f}} l_1 \left(\frac{\mathrm{d}i_a(t)}{\mathrm{d}t} - \frac{\mathrm{d}i_b(t)}{\mathrm{d}t} \right), \tag{4.33}$$

where $v_{ab}(t)$ is the line-to-line voltage between phase A and phase B; i_a and i_b are the currents of phase A and phase B at the relay position, respectively.

In this case, (4.33) can be simplified to:

$$v_{ab}(t) = \left[H_{1ab}(t)\ H_{2ab}(t) \right] \begin{bmatrix} R_{\mathrm{f}}' \\ D_{\mathrm{f}} \end{bmatrix}, \tag{4.34}$$

where

$$H_{1ab}(t) = 0,$$

and

$$H_{2ab}(t) = r_1 (i_a(t) - i_b(t)) + l_1 \frac{\mathrm{d}i_a(t)}{\mathrm{d}t} - l_1 \frac{\mathrm{d}i_b(t)}{\mathrm{d}t}.$$

Using numerical derivative instead, $H_{2ab}(t)$ can be transformed with a first-order derivative and becomes

$$H_{2ab}(n) = r_1 (i_a(n) - i_b(n)) + l_1 \frac{i_a(n) - i_a(n-1)}{\Delta t} - l_1 \frac{i_b(n) - i_b(n-1)}{\Delta t}. \tag{4.35}$$

4.2.3.3 Estimation of the Fault Distance

The fault distance can be calculated using the recursive least squares (RLS) method represented in a discrete-time form. At time $k+1$, the unknown parameters, R_{f}' and D_{f}, can be estimated from their values at instant k using the following recursive formula:

$$[P_{k+1}] = [P_k] + [G_k] \cdot \left[v(k) - [H_k]\,[P_k]^T \right], \tag{4.36}$$

where

$$[P_k] = \begin{bmatrix} R_{\mathrm{f}}'(k) \\ D_{\mathrm{f}}(k) \end{bmatrix},$$

$$[G_k] = \begin{bmatrix} G_1(k) \\ G_2(k) \end{bmatrix},$$

and

$$[H_k] = \begin{bmatrix} H_1(k) \\ H_2(k) \end{bmatrix}.$$

For a phase-A-ground fault, $v(k) = v_a(k)$, $H_1(k) = H_{1a}(k)$ and $H_2(k) = H_{2a}(k)$. In (4.36), G_k is the gain vector, which is obtained by:

$$[G_k] = \frac{[J_k]\,[H_k]}{I + [H_k]^T\,[J_k]\,[H_k]}. \tag{4.37}$$

where $[J_k]$ is a 2×2 matrix and is updated recursively by:

$$[J_{k+1}] = [J_k]\left[I - [G_k]\,[H_k]^T\right]. \tag{4.38}$$

For the first step, J_0 is initiated to be a diagonal matrix in which each element is assigned a large positive value. The estimated values of the ground-fault resistance and distance are updated step by step with new instantaneous currents and voltages sampled for calculation.

4.3 Implementation of ADRA

4.3.1 Calculation of the Fault Distance by RLS

The fault distance and impedance can be estimated by the RLS technique. The estimation process is initialised when the input current reaches a steady state after a fault occurs. The MFD output, $\triangle D(k)$, and counter $C(k)$ are used as indicators of the state of currents. If $C(k) > 0$ or $\triangle D(k) > M$, the fault current is in a transient state. In this case, the fault current contains high frequency components, which significantly influence the accuracy of the fault distance estimation. Therefore, the estimation process should be frozen during this transient period, until both $C(k) = 0$ and $\triangle D(k) \leq M$ are satisfied.

RLS algorithm (4.36) works based on the instantaneous current and voltage samples, i.e. six quantities sampled at the same time instant, which are defined as a group of samples. The number of groups involved in the estimation process is denoted as N_{w}, which increases step by step during the estimation process. At each step of recursive calculation, MFD will be used, based on each new group of samples, to check whether the sampled quantities are in a transient state before the algorithm is executed. If so, the calculation of $R'_{\mathrm{f}}(k)$ and $D_{\mathrm{f}}(k)$ should be invalid. The detailed process of the fault distance estimation is illustrated in Fig. 4.4.

The estimation process stops when a disturbance is detected by MFD and continues afterwards, which makes the scheme robust against noise. It should be mentioned that although the RLS algorithm represented by (4.36) is in a generic form for any type of fault, $v(k)$ and $H(k)$ in (4.36) should be assigned with the relevant currents and voltages in link with the fault type. The accuracy of $R'_{\mathrm{f}}(k)$ and $D_{\mathrm{f}}(k)$

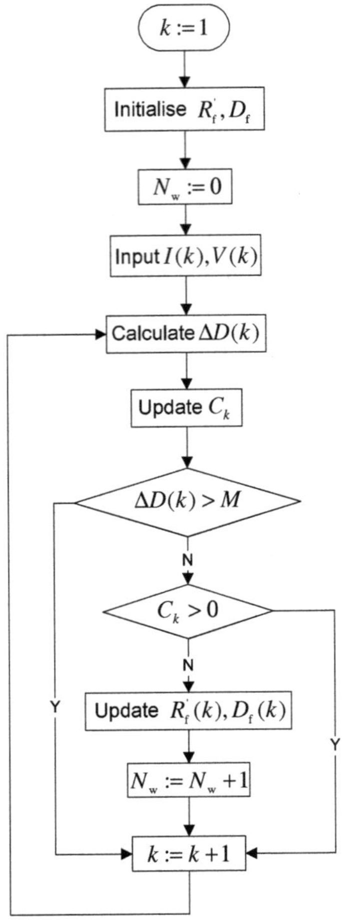

Fig. 4.4 The process of fault distance estimation

improves as N_w increases. The convergence of $R_f'(k)$ and $D_f(k)$ depends on the convergence of the gain, G_k. From our experience, the RLS algorithm converges rapidly. Once it starts, it only requires a few groups of samples, i.e. a few steps, to reach the steady values of $R_f'(k)$ and $D_f(k)$. In this case, the calculation could be completed within a very short period of time if the sampling rate of the distance relay is fast enough. To be conservative, we take a fixed period of time, such as a quarter of the fundamental cycle, rather than a fixed number of steps, to monitor the convergence of the RLS algorithm no matter what the sampling rate is. Let N_s denote the sampling rate per fundamental cycle. If $N_s = 64$, $N_s/4$ would be enough for $R_f'(k)$ and $D_f(k)$ to converge, within a quarter of the fundamental cycle.

4.3.2 Settings for the Variable Tripping Zone

In order for the distance relay to have an accurate and reliable relying action, the accuracy of estimated $R'_f(k)$ and $D_f(k)$ is considered to determine a tripping zone. As the accuracy of estimation is improved step by step during the estimation process as N_w increases, the tripping zone is set to be variable. Figure 4.5 illustrates the relationship between the estimation accuracy and the setting of the tripping zone.

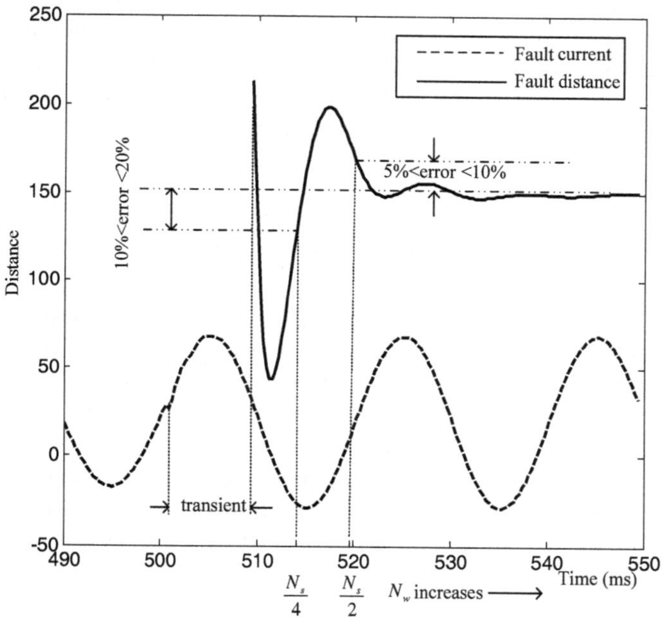

Fig. 4.5 The setting of the adaptive tripping zone. The *dotted* and *solid curves* denote the fault current waveform and the estimated fault distance, respectively

From Fig. 4.5, it can be seen that the maximum reach of the tripping zone, L_{set}, increases as N_w increases. The relay makes a tripping decision if $N_w > \lfloor N_s/4 \rfloor$, which means the relay is blocked as long as the estimation process is shorter than a quarter of the fundamental cycle. If $N_w = \lfloor N_s/4 \rfloor$, then the maximal reach of the tripping zone, L_{set}, is set to 65% of L, the length of the transmission line. The maximal reach of the tripping zone increases as N_w increases, until it reaches about 80% of the length of the transmission line for $N_w = N_s$. Finally, the tripping zone is set to its maximum with a reach of 90% for $N_w > N_s$. It should be mentioned that the closer the fault location to the relay position, the less accuracy of estimation is required. Moreover, the estimation accuracy is also related to different types of faults. Considering these factors, we adopt a more conservative scheme to configure the setting of tripping zones by determining L_{set} as follows:

$$L_{\text{set}} = \begin{cases} 0.65L, & \lfloor \frac{N_s}{4} \rfloor \le N_w \le \lfloor \frac{N_s}{2} \rfloor \\ 0.80L, & \lfloor \frac{N_s}{2} \rfloor < N_w \le N_s \\ 0.90L, & N_w > N_s \end{cases} . \tag{4.39}$$

Consequently, a fault location close to the relay position can be detected within half a cycle after fault inception, while the tripping time for faults around the remote end is comparable to that of full-cycle FT-based algorithms. In the mean time, the reliability of ADRA is guaranteed by adopting a variable tripping zone.

Since the fault distance is estimated in a recursive manner, ADRA is as robust to distorted input signals as standard full-cycle FT-based algorithms. Furthermore, the simultaneous estimation of ground-fault resistance improves the reliability of the scheme for faults occurring around the remote line end.

4.4 Simulation Studies and Results of ADRA

In order to evaluate the performance of ADRA, a simple transmission network has been simulated using the electromagnetic transient computer program PSCAD/EMTDC. The model used for the simulation studies is shown in Fig. 4.6. In this model, the 230 kV transmission line has a length of 240 km, with a positive-sequence impedance $0.0343 + j0.4188\ \Omega$ and zero-sequence impedance $0.2913 + j1.1566\ \Omega$ per kilometre. The equivalent impedances of power sources in zone A and zone B are $5 + j27.7\ \Omega$ and $0.6 + j9.3\ \Omega$, respectively. The voltages of power sources S1 and S2 in zone A and zone B are 230 kV and $230\angle\delta$ kV, respectively, where δ is the load angle in degrees.

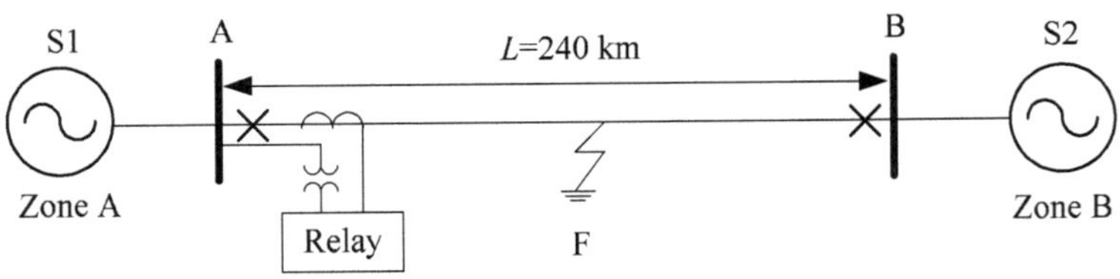

Fig. 4.6 A transmission line model

A distance relay is located at busbar A. The relay measures phase currents and three line-to-neutral voltages of the transmission line at the relay position. The six signals are transformed by CTs and VTs, respectively. Each CT is designed with a transfer ratio of 100 : 1 and each VT has a transfer ratio of 100 : 1 as well. Both CTs and VTs are simulated based on the Lucas model [94]. The measured continuous analogue signals are then sampled at a frequency of 3200 Hz under a fundamental frequency of 50 Hz, which means that the currents and voltages are sampled at the rate of 64 points per fundamental cycle.

4.4.1 Ground Faults with Ground-Fault Resistance

ADRA has been tested for phase-A-ground faults. The relay is installed at busbar A and protects the transmission line up to busbar B. The tests conducted are concerned with ground faults, including different ground-fault resistances at various fault locations. The effect of different power angles, δ, influencing the distance measurement for ground faults, is also considered in the tests.

Figure 4.7 illustrates a phase-A-ground fault applied at 120 km from the relay at busbar A. The fault occurs at time instant 500 ms, with a load angle $\delta = -20°$ and a ground-fault resistance $R_f = 20\ \Omega$. The phase currents I_A, I_B, I_C and zero-sequence current I_0 are measured and shown in Fig. 4.7a. The MFD output, ΔD_i, is calculated for each of the currents and plotted in Fig. 4.7b. Threshold M of ΔD_i is pre-set to 0.5 for all the cases concerned.

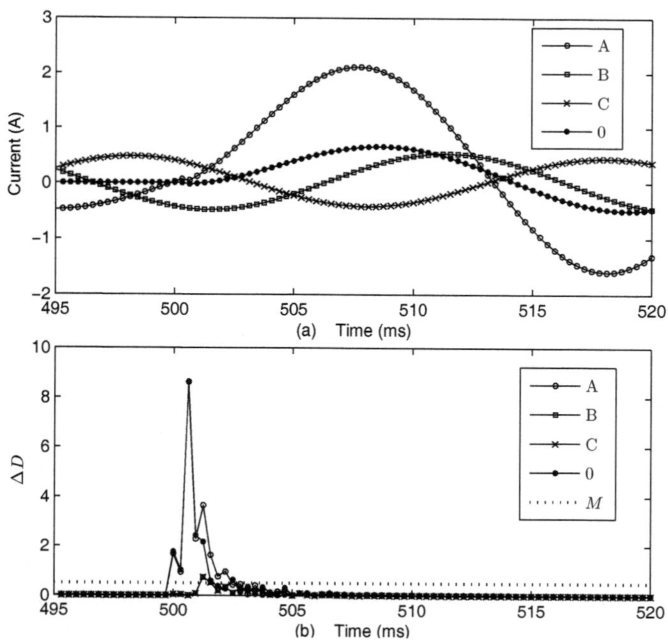

Fig. 4.7 A single-phase-to-ground fault applied at 120 km from busbar A. **a** The three phases and zero-sequence currents. **b** The MFD outputs of the four currents

The norm of the MFD output, E_i, is calculated for the purpose of fault classification. In the estimation of E_i, a window of a size N_D ($N_D = 5$) containing ΔD_i

is used. The norms E_i, corresponding to the four currents, are given in Fig. 4.8a. Threshold ε is set to be 4 for fault classification. From Fig. 4.8a, it can be seen that two norms obtained from currents I_A and I_0 exceed the threshold during the fault period, which indicates the occurrence of a phase-A-ground fault. Furthermore, the counter C is used for the purpose of monitoring the values of MFD output, ΔD, and is also compared with threshold C_{set}, where C_{set} is set to be 2. The counter for the MFD output of phase A is given in Fig. 4.8b, which indicates the transient period of the current I_A after the fault occurs.

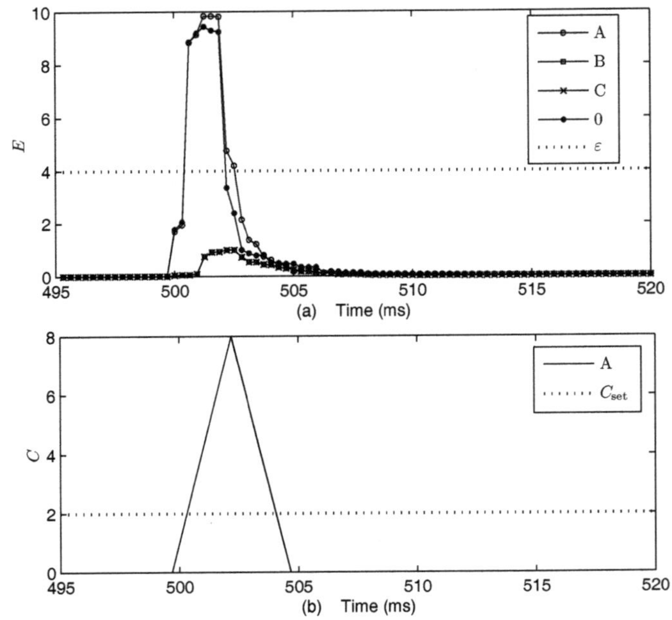

Fig. 4.8 Fault detection and classification. **a** The norms of the MFD output for the fault currents. **b** The counter of MFD for phase A

To estimate the fault distance and ground-fault resistance and to avoid using the samples of currents and voltages during the transient period of faults, the MFD output and the counter are employed as indicators. In this case, the recursive calculation of the fault distance and the ground-fault resistance begins at time instant 504.69 ms, by monitoring $\Delta D \leq M$ and $C = 0$ for the fault current sampled at that specific time. The estimation process continues for two fundamental cycles until time instant 544.69 ms. The estimated values of the fault distance and the ground-fault resistance are plotted in Figs. 4.9a and b, respectively. According to the setting of the variable tripping zone for the distance protection, the relay makes a tripping decision at time instant 509.69 ms, which is a quarter of the fundamental cycle, i.e. 5 ms after the estimation process starts. This indicates that the relay can make a tripping decision at time instant 9.69 ms after the fault occurs, which is much shorter than the tripping time of the relay using conventional distance relaying algorithms.

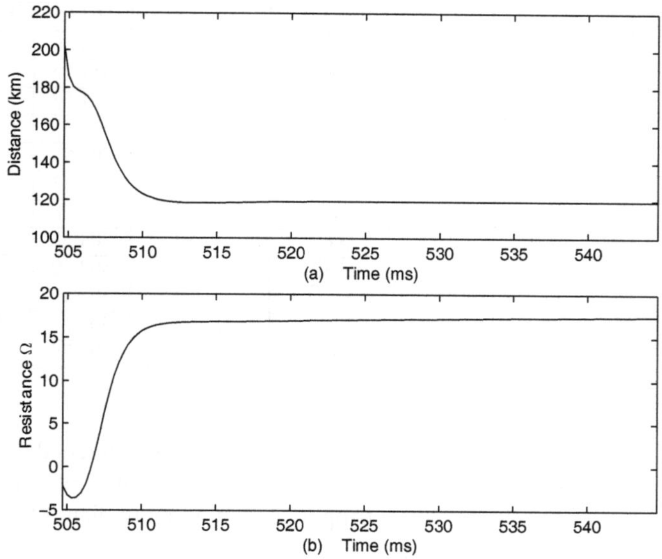

Fig. 4.9 Estimation results of the fault distance and the ground-fault resistance. **a** The estimated fault distance. **b** The estimated ground-fault resistance

Similar cases are investigated with consideration of various fault locations, ground-fault resistances and load angles. The results of the estimated fault distance, the ground-fault resistances and the tripping time of the relay are listed in Table 4.1. To show the period of the estimation process, the starting time of the RLS algorithm is also listed in Table 4.1. From Table 4.1, it can be seen that the performance of ADRA is satisfactory and the faults occurring at different locations and the associated distance impedances are correctly identified. For any faults occurring within the zone of 192 km, i.e. 80% of the length of the transmission line, the tripping time of ADRA would be in the range from 6.25 ms to 19.06 ms, which is no more than the duration of a fundamental cycle. This implies that ADRA is able to take the tripping actions in less time in comparison with conventional distance relays that usually require current and voltage samples obtained from a whole period of a fundamental cycle to make a decision.

The estimation results of the fault distance and the ground-fault resistance listed in Table 4.1 are reasonably accurate, with a maximal error of 2% of the actual value for single-phase-to-ground faults. This accuracy enables ADRA to protect a wider zone of up to 90% or more of the length of the transmission line in its first protection zone. For the purpose of reliability, if a fault occurs outside 80% of the length of the transmission line, the recursive estimation algorithm used in ADRA will repeat until

Table 4.1 Summary of cases for single-phase-to-ground faults

Fault distance (km)	Fault resistance (Ω)	Load angle (°)	Estimated distance (km)	Estimated resistance (Ω)	Detection time (ms)	Tripping time (ms)
30	0	-15	29.63	-1.63	502.18	507.18
90	20	-30	92.57	15.48	504.38	509.38
150	50	10	155.84	45.63	504.06	513.13
180	100	-15	175.32	96.47	506.56	516.56
210	0	-30	209.25	-1.66	511.25	531.25
30	20	10	33.44	16.36	501.88	506.88
90	50	-15	94.97	45.85	504.38	509.38
150	100	-30	147.44	96.95	505.94	510.94
180	0	10	184.00	-2.41	505.93	515.93
210	20	-15	206.05	20.53	509.38	529.38
30	50	-30	33.68	46.83	504.38	509.38
90	100	10	93.82	94.24	502.50	507.50
150	0	-15	151.71	-3.00	507.19	512.19
180	20	-30	177.71	17.61	509.69	519.69
210	50	10	206.39	54.07	506.25	526.25

$N_w > N_s$, i.e. the whole estimation process is longer than a fundamental cycle in this case. As shown in Table 4.1, for fault cases with a fault distance of 210 km from busbar A, the relay can reliably take a tripping action no more than 30.63 ms after fault occurrence.

4.4.2 Other Types of Faults

The simulation studies are extended for phase-to-phase, double-phase-to-ground, and three-phase faults. Figure 4.10 illustrates the fault currents and the MFD output caused by a phase-A-B fault with a fault distance at 80 km from the relay position. The load angle in this case is $-15°$. The phase currents and the zero-sequence current are shown in Fig. 4.10a and their MFD outputs, ΔD, are calculated and plotted in Fig. 4.10b.

Based on the results of the MFD outputs, the norms of these outputs, in respect to the phase and zero-sequence currents, are calculated, and their values are plotted against time as shown in Fig. 4.11a. In this case, ADRA is able to recognise the fault as a phase-A-B fault based upon the norms, since the norms corresponding to phase A and phase B are larger than threshold ε, while the other norms are not.

According to the MFD output of phase A shown in Fig. 4.11a and the values of the counter of phase A shown in Fig. 4.11b, it can be seen from Fig. 4.12 that the estimation process begins at time instant 504.38 ms and a tripping action is taken after a quarter of the fundamental cycle at 509.38 ms. At this time instant, the fault distance is estimated as 77.26 km with an error smaller than 1.2%, which guarantees the reliability and accuracy of the distance protection.

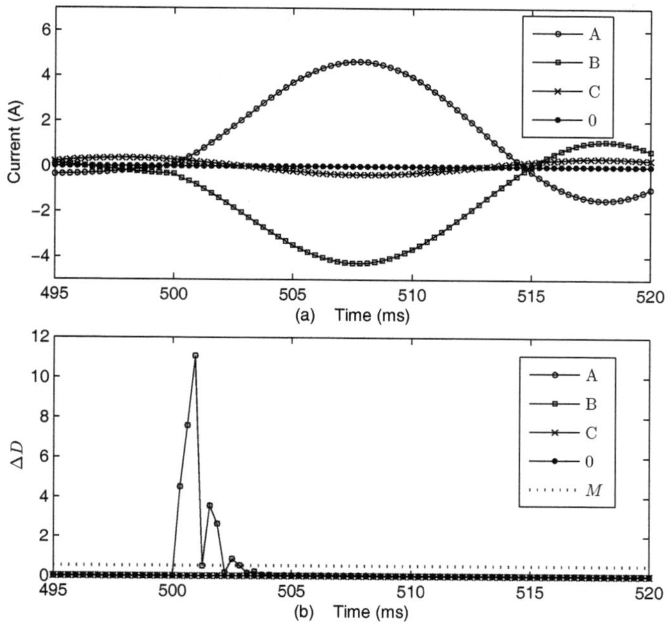

Fig. 4.10 A phase-A-B fault applied at 80 km from busbar A. **a** The three phases and zero-sequence currents. **b** The MFD outputs of the four currents

Other fault cases are simulated under different fault locations and load angles. The results obtained from each case, including the estimated fault distance and relay tripping time, are given in Table 4.2. For the cases that have a fault distance within 80% of the transmission line (< 192 km), as listed in Table 4.2, ADRA can take a tripping action within a range from 6.88 ms to 19.06 ms, which indicates a significant improvement in speed compared with conventional distance relaying algorithms. Faults that have a distance of 210 km from the relay position are also considered. Such a fault distance is not within the first protection zone that is usually covered by the conventional distance relays. They are commonly detected and tripped with a time delay around 0.5 s as a backup protection strategy. In ADRA, the first protection zone is extended, since the fault distance can be measured more accurately. In Table 4.2, the errors of the estimated distances for fault cases occurring 210 km from the relay position are within 2%, which indicates that the faults take place inside the transmission line. In order to maintain the reliability of locating long distance faults, ADRA continues to calculate the fault distance for one cycle and a tripping decision will be issued after 35 ms of the fault occurrence.

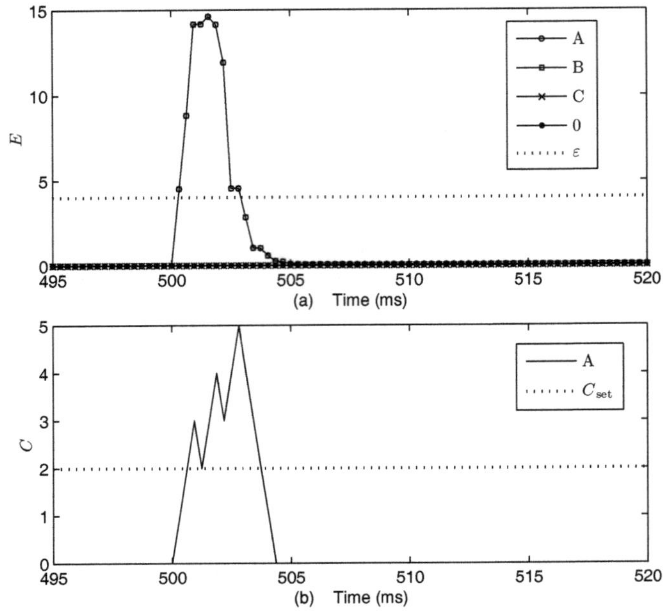

Fig. 4.11 Fault detection and classification. **a** The norms of the MFD outputs for the fault currents. **b** The counter of MFD for phase A

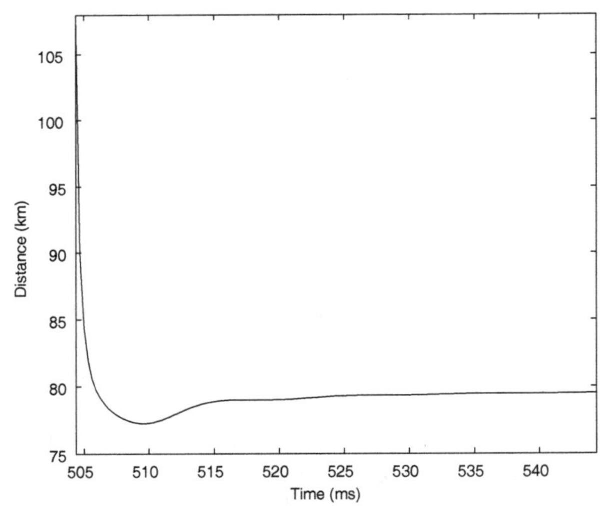

Fig. 4.12 Fault distance estimation results

Table 4.2 Summary of cases for other fault types

Fault Type	Fault distance (km)	Load angle (°)	Estimated distance (km)	Detection time (ms)	Tripping time (ms)
A-B	30	-15	29.66	503.12	508.12
B-C-G	90	-30	90.93	505.63	510.63
A-B-C	150	10	155.93	507.19	512.50
A-B-C-G	180	-15	183.91	509.06	519.06
B-C	210	-30	213.68	515.63	535.63
C-A-G	30	10	30.10	505.00	510.00
A-B-C	90	-15	87.53	503.75	508.75
A-B-C-G	150	-30	155.78	508.44	515.00
C-A	180	10	184.19	510.94	515.31
A-B-G	210	-15	213.53	510.00	530.00
A-B-C	30	-30	29.84	502.50	507.50
A-B-C-G	90	10	89.40	501.88	506.88

4.4.3 External Faults

The performance of ADRA has also been tested for faults outside the first protection zone of the relay ($> 0.9L$). Figure 4.13 shows the results of the estimated distance for a double-phase-to-ground fault (phase-A-B-ground) simulated on the transmission line at a distance of 238 km from busbar A. This fault simulation is carried out with a ground-fault resistance of 10 Ω and a load angle of 10°.

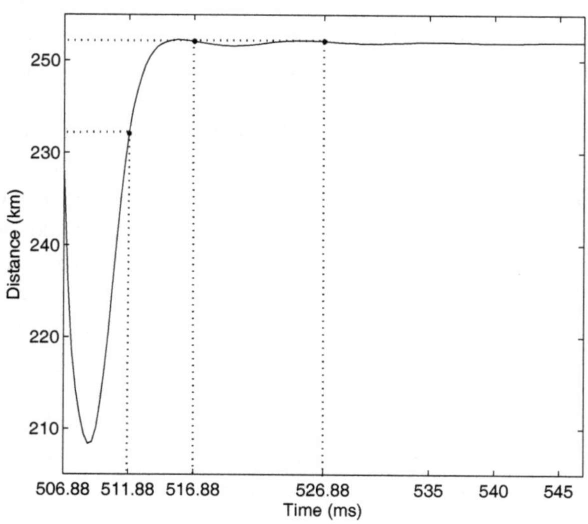

Fig. 4.13 The estimated fault distance for an external fault

As shown in Fig. 4.13, the fault is detected and the distance estimation process begins at time instant 506.88 ms. After a quarter of the fundamental cycle at 511.88 ms, the result of the distance estimation is 232.07 km, which is outside the tripping zone of $0.65L$ (156 km). Hence, the distance estimation process continues for half a cycle and then to a whole cycle. The estimation results are 242.08 km and 242.10 km, respectively. It is observed that the estimated distance converges around 242.10 km, which is outside the operating region of the relay at $0.8L$ (192 km) and $0.9L$ (216 km). This leads to the decision that the fault is outside the first protection zone of the relay. Thus, no tripping command will be issued.

Results from the simulation studies reveal that ADRA can guarantee a fast tripping time for near faults and high reliability for remote faults close to the reach boundary. The algorithm provides a faster relaying response and higher reliability than conventional DFT-based distance protection methods.

4.5 Protection of a Double Circuit Transmission Line

Double circuit transmission lines are widely installed in modern electric power networks, thanks to their significant economic and environmental advantages [9]. However, it is difficult to classify the fault types in such lines, because of the effect of mutual coupling between the two circuits. Under a fault condition, the transient signals induced on the two circuits are quite different from those encountered on a single circuit line. The fault phases on one circuit have a coupling effect on the phases on the other healthy circuit, which may make the healthy circuit to be wrongly diagnosed under certain fault conditions [63]. Thus, conventional fault classifiers, based on power frequency measurement, suffer from deficiencies due to the influence of mutual coupling between the two circuits, and cannot be applied to fault classification for protection of a double circuit transmission line.

Approaches using ANN have been employed for fault classification of a double circuit transmission line [5, 146]. In [5], a self-organisation map-based network was used to correctly identify the fault phases in spite of the presence of the highly variable mutual coupling between the two circuits. However, to make this technique feasible, much computational effort has to be made in the pre-processing and training stages. Another approach using the travelling wave theory has been introduced to perform the selection of fault phases of a double circuit transmission line [76]. Although the information of travelling waveforms can potentially enable an accurate fault phase selection within half a fundamental cycle, this approach leads to increased hardware requirements for the high sampling frequency. Meanwhile, sampled at such a high frequency, the travelling waveforms cannot be separated from interference noise. Thus, an alternative fault classifier is desired.

MGW can be employed to design a fault phase selector. It is used to analyse the six phase currents, together with a ground mode current obtained by a modal transformation. By extracting the fault components from the currents using MGW, the fault phases can be identified rapidly and reliably. Simulation studies are conducted on a double circuit transmission model. The test results are analysed and show that the MGW-based fault phase selector is effective in accurately classifying the types of faults occurring on the double circuit transmission line.

4.6 The Fault Phase Selector

In our approach, MGW is used to depress the steady-state components and obtain the gradient information from fault currents. The construction of MGW was presented in Sect. 2.4.3.2. The six phases on the two circuits of a double circuit transmission line are mutually coupled due to the mutual inductance between adjacent circuits of the line. Using an appropriate modal transformation, the coupled phases can be decoupled to obtain six independent phases. This process is known as modular analysis, which can be extended from a three-phase transmission line to the six-phase double circuit line [57]. In this section, a new modal transformation is adopted. The six sequence components are not only symmetrical to the three phases in a single circuit, but also to all six phases on the double circuit transmission line. The process of the modal transformation is described in detail as follows.

Assuming that $[\dot{U}_{\mathrm{I\,II}}]$, $[\dot{I}_{\mathrm{I\,II}}]$, and $[Z]$ represent the matrices of the voltage, current and impedance of the double circuit of transmission line, we have:

$$[\dot{U}_{\mathrm{I\,II}}] = [Z][\dot{I}_{\mathrm{I\,II}}]. \tag{4.40}$$

The above equation can be written in detail as:

$$\begin{bmatrix} \dot{U}_{\mathrm{I\,A}} \\ \dot{U}_{\mathrm{I\,B}} \\ \dot{U}_{\mathrm{I\,C}} \\ \dot{U}_{\mathrm{II\,A}} \\ \dot{U}_{\mathrm{II\,B}} \\ \dot{U}_{\mathrm{II\,C}} \end{bmatrix} = \begin{bmatrix} Z_s & Z_m & Z_m & Z_m' & Z_m' & Z_m' \\ Z_m & Z_s & Z_m & Z_m' & Z_m' & Z_m' \\ Z_m & Z_m & Z_s & Z_m' & Z_m' & Z_m' \\ Z_m' & Z_m' & Z_m' & Z_s & Z_m & Z_m \\ Z_m' & Z_m' & Z_m' & Z_m & Z_s & Z_m \\ Z_m' & Z_m' & Z_m' & Z_m & Z_m & Z_s \end{bmatrix} \begin{bmatrix} \dot{I}_{\mathrm{I\,A}} \\ \dot{I}_{\mathrm{I\,B}} \\ \dot{I}_{\mathrm{I\,C}} \\ \dot{I}_{\mathrm{II\,A}} \\ \dot{I}_{\mathrm{II\,B}} \\ \dot{I}_{\mathrm{II\,C}} \end{bmatrix},$$

where Z_s is the self impedance for each phase, Z_m is the mutual impedance between phases in one circuit, and Z_m' is the mutual impedance of two phases between the two circuits. $\dot{U}_{\mathrm{I\,A}}, \dot{U}_{\mathrm{I\,B}}, \dot{U}_{\mathrm{I\,C}}$ are three phase voltages of the first circuit, with a subscript of I; $\dot{U}_{\mathrm{II\,A}}, \dot{U}_{\mathrm{II\,B}}, \dot{U}_{\mathrm{II\,C}}$ denote the three phase voltages of the second circuit, with a subscript of II. The same denotation applies to currents for each phase.

To decouple the six phase components, the currents are divided into two parts: positive-sequence and negative-sequence components, which are represented by $[\dot{I}_{\mathrm{T}}]$ and $[\dot{I}_{\mathrm{F}}]$ respectively. Define a transformation matrix $[P]$ to be:

$$[P] = \begin{bmatrix} 1 & 0 & 0 & 1 & 0 & 0 \\ 0 & 1 & 0 & 0 & 1 & 0 \\ 0 & 0 & 1 & 0 & 0 & 1 \\ 1 & 0 & 0 & -1 & 0 & 0 \\ 0 & 1 & 0 & 0 & -1 & 0 \\ 0 & 0 & 1 & 0 & 0 & -1 \end{bmatrix}, \tag{4.41}$$

$$[P]^{-1} = \frac{1}{2}[P], \tag{4.42}$$

then the phase voltages and currents are transformed to:

$$[\dot{U}_{\mathrm{I\,II}}] = [P][\dot{U}_{\mathrm{TF}}], \quad [\dot{I}_{\mathrm{I\,II}}] = [P][\dot{I}_{\mathrm{TF}}], \tag{4.43}$$

where $[\dot{U}_{\mathrm{I\,II}}] = [\dot{U}_{\mathrm{I}}, \dot{U}_{\mathrm{II}}]^{\mathrm{T}}$, $[\dot{I}_{\mathrm{I\,II}}] = [\dot{I}_{\mathrm{I}}, \dot{I}_{\mathrm{II}}]^{\mathrm{T}}$, $[\dot{U}_{\mathrm{TF}}] = [\dot{U}_{\mathrm{T}}, \dot{U}_{\mathrm{F}}]^{\mathrm{T}}$, and $[\dot{I}_{\mathrm{TF}}] = [\dot{I}_{\mathrm{T}}, \dot{I}_{\mathrm{F}}]^{\mathrm{T}}$. Replacing $[\dot{U}_{\mathrm{I\,II}}]$ and $[\dot{I}_{\mathrm{I\,II}}]$ by $[\dot{U}_{\mathrm{TF}}]$ and $[\dot{I}_{\mathrm{TF}}]$ in (4.40), we obtain:

$$[\dot{U}_{\mathrm{TF}}] = [P]^{-1}[Z][P][\dot{I}_{\mathrm{TF}}]. \tag{4.44}$$

The impedance matrix is converted to:

$$[P]^{-1}[Z][P] = \begin{bmatrix} Z_{\mathrm{T}} & 0 \\ 0 & Z_{\mathrm{F}} \end{bmatrix}, \tag{4.45}$$

where

$$[Z_{\mathrm{T}}] = \begin{bmatrix} Z_s + Z'_m & Z_m + Z'_m & Z_m + Z'_m \\ Z_m + Z'_m & Z_s + Z'_m & Z_m + Z'_m \\ Z_m + Z'_m & Z_m + Z'_m & Z_s + Z'_m \end{bmatrix} \tag{4.46}$$

and

$$[Z_{\mathrm{F}}] = \begin{bmatrix} Z_s - Z'_m & Z_m - Z'_m & Z_m - Z'_m \\ Z_m - Z'_m & Z_s - Z'_m & Z_m - Z'_m \\ Z_m - Z'_m & Z_m - Z'_m & Z_s - Z'_m \end{bmatrix}. \tag{4.47}$$

Based on the above transformation, (4.40) is split into two parts:

$$[\dot{U}_{\mathrm{T}}] = [Z_{\mathrm{T}}][\dot{I}_{\mathrm{T}}], \quad [\dot{U}_{\mathrm{F}}] = [Z_{\mathrm{F}}][\dot{I}_{\mathrm{F}}]. \tag{4.48}$$

The modal transformation is applied to the two parts, respectively, to eliminate the mutual impedance in Z_{T} and Z_{F}. Here, a symmetrical component transformation is adopted, which is described as:

$$[S] = \frac{1}{3}\begin{bmatrix} 1 & a & a^2 \\ 1 & a^2 & a \\ 1 & 1 & 1 \end{bmatrix}, \tag{4.49}$$

where $a = e^{j120^\circ}$. Hence, the two decoupled parts are obtained and denoted as:

$$\begin{bmatrix} \dot{U}_{T0} \\ \dot{U}_{T1} \\ \dot{U}_{T2} \end{bmatrix} = \begin{bmatrix} Z_{T0} & 0 & 0 \\ 0 & Z_{T1} & 0 \\ 0 & 0 & Z_{T2} \end{bmatrix} \begin{bmatrix} \dot{I}_{T0} \\ \dot{I}_{T1} \\ \dot{I}_{T2} \end{bmatrix} \quad (4.50)$$

and

$$\begin{bmatrix} \dot{U}_{F0} \\ \dot{U}_{F1} \\ \dot{U}_{F2} \end{bmatrix} = \begin{bmatrix} Z_{F0} & 0 & 0 \\ 0 & Z_{F1} & 0 \\ 0 & 0 & Z_{F2} \end{bmatrix} \begin{bmatrix} \dot{I}_{F0} \\ \dot{I}_{F1} \\ \dot{I}_{F2} \end{bmatrix}, \quad (4.51)$$

respectively. Z_{T0}, Z_{T1}, Z_{T2} are zero-sequence, positive-sequence and negative-sequence impedances of the first circuit; and Z_{F0}, Z_{F1}, Z_{F2} are zero-sequence, positive-sequence and negative-sequence impedances of the second circuit. Combining the transformation matrices $[S]$ and $[P]$, a new transformation matrix for the six-phase double circuit line can be represented as:

$$[M] = \begin{bmatrix} 1 & 1 & 1 & 1 & 1 & 1 \\ 1 & 1 & a^2 & a^2 & a & a \\ 1 & 1 & a & a & a^2 & a^2 \\ 1 & -1 & 1 & -1 & 1 & -1 \\ 1 & -1 & a^2 & -a^2 & a & -a \\ 1 & -1 & a & -a & a^2 & -a^2 \end{bmatrix}. \quad (4.52)$$

Based on the transformation, (4.50) and (4.51) are integrated into one equation as:

$$\begin{bmatrix} \dot{U}_{T0} \\ \dot{U}_{T1} \\ \dot{U}_{T2} \\ \dot{U}_{F0} \\ \dot{U}_{F1} \\ \dot{U}_{F2} \end{bmatrix} = \begin{bmatrix} Z_{T0} & 0 & 0 & 0 & 0 & 0 \\ 0 & Z_{T1} & 0 & 0 & 0 & 0 \\ 0 & 0 & Z_{T2} & 0 & 0 & 0 \\ 0 & 0 & 0 & Z_{F0} & 0 & 0 \\ 0 & 0 & 0 & 0 & Z_{F1} & 0 \\ 0 & 0 & 0 & 0 & 0 & Z_{F1} \end{bmatrix} \begin{bmatrix} \dot{I}_{T0} \\ \dot{I}_{T1} \\ \dot{I}_{T2} \\ \dot{I}_{F0} \\ \dot{I}_{F1} \\ \dot{I}_{F2} \end{bmatrix},$$

which can also be represented as:

$$[\dot{U}_{TF}] = [Z_{TF}][\dot{I}_{TF}], \quad (4.53)$$

where $[Z_{TF}] = [M]^{-1}[Z][M]$, $[\dot{U}_{TF}] = [M]^{-1}[\dot{U}_{\mathrm{I\,II}}]$ and $[\dot{I}_{TF}] = [M]^{-1}[\dot{I}_{\mathrm{I\,II}}]$.

The use of the modal transformation aids the fault type identification. The signals in the ground mode help discriminate whether a fault connects with earth. The presence of a small but non-null signal in the ground mode indicates that there exists a ground fault. According to the modal transformation, the ground mode current I_{T0} is calculated as:

$$I_{T0} = \frac{1}{6}(I_{\mathrm{I\,A}} + I_{\mathrm{I\,B}} + I_{\mathrm{I\,C}} + I_{\mathrm{II\,A}} + I_{\mathrm{II\,B}} + I_{\mathrm{II\,C}}), \quad (4.54)$$

where $I_{\mathrm{I\,A}}, \ldots, I_{\mathrm{II\,C}}$ denote the currents of six different phases, respectively. The ground mode current and the six phase currents are composed as the input signals that can be processed by MGW. The transient features can be extracted from the currents by MGW and represented as detail signals that can provide sufficient information for fault phase selection. For the MGW detail design, the reader is referred to Sect. 2.4.3.2.

In total, the seven current signals are fed to MGW and their detail signals can be obtained, which are denoted as ΔD_i, and $i = 1,2,\ldots,7$. Here, $i = 1$ corresponds to the ground mode current, while $2,\ldots,7$ indicate the six phase currents, respectively. The length of MGW contains three samples according to (2.62). The norms of ΔD_i are calculated to measure the amount of energy contained in D_i. The norms, denoted as E_i, can be calculated as follows:

$$E_i = \sqrt{\sum_{k=1}^{N_\mathrm{D}} \Delta D_i^2(k)}, \quad i = 1,2,\ldots,7, \tag{4.55}$$

where N_D is the number of samples in a sliding data window. Its length equals a cycle at a fundamental frequency of 50 Hz. The subscript i refers to the phase number. Similarly to (4.18), the discrimination components can be derived as follows:

$$L_i = \begin{cases} 1 & E_i > \varepsilon \\ 0 & E_i < \varepsilon \end{cases}, \tag{4.56}$$

where ε is a threshold that can be determined by pre-calculation for each individual transmission circuit, and L_i are logic values that are used as discriminant components for fault detection and fault phase selection. The phase selection is obtained by the comparison of the norms of all the seven input currents, because the norms of fault phases are much higher than those of healthy phases. If a norm exceeds a pre-determined threshold, its corresponding phase is recognised as the fault phase. The process of fault phase selection is illustrated in Fig. 4.14.

4.7 Simulation Results and Discussions on Fault Phase Selection

The MGW-based phase selector has been evaluated on a power system model, shown in Fig. 4.15, which is simulated using the PSCAD/EMDTC program. It consists of a 400 kV double circuit transmission line, which is connected to power sources at both ends. The double circuit transmission line is 200 km long, and the configuration of its geometry components is illustrated in Fig. 4.16. In Fig. 4.16, G1 and G2 are two ground wires. C1, C2 and C3 are the three phase of wires of the first circuit, while C4, C5 and C6 are the three phase of wires for the second circuit. 'N1Tower', 'N1Conductor', and 'N1Groundline' denote the names of the tower, conductors and ground wires, respectively. The double circuit transmission line has a zero-sequence impedance $Z_{L0} = 0.85 + j5.43\ \Omega$ and a positive-sequence impedance $Z_{L1} = 0.4 + j2.8\ \Omega$. The power source voltages are $E_{\mathrm{s}1} = 400$ kV and $E_{\mathrm{s}2} = 400\angle\delta$ kV, where δ is the load angle in degrees.

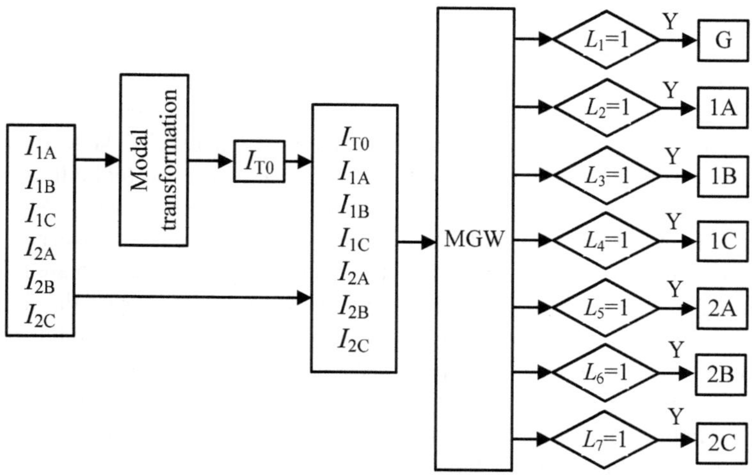

Fig. 4.14 Block diagram of the phase selector

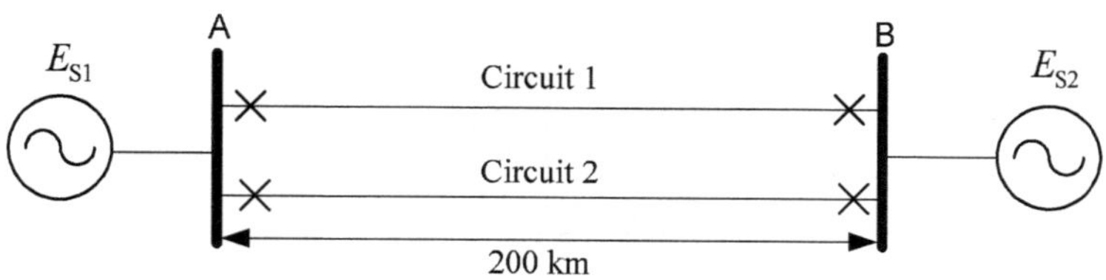

Fig. 4.15 A double circuit transmission line model

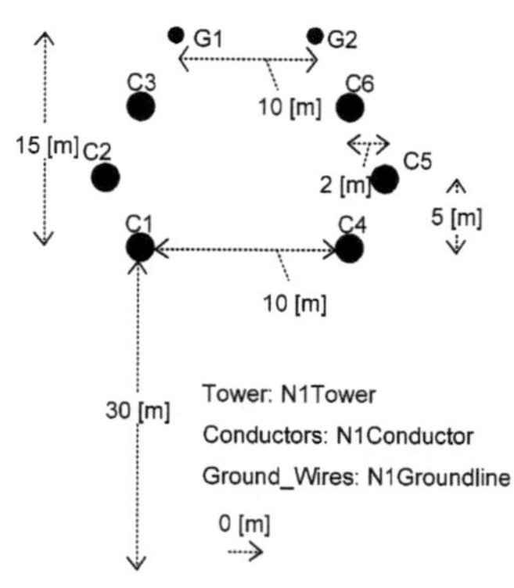

Fig. 4.16 Geometry components of the double circuit transmission line

Two relays are placed at both ends of the double circuit line under protection to ensure fault isolation. All six phase currents at the end of the double circuit line are measured using 15A/5A CTs. The signals are then sampled at a sampling frequency of 1600 Hz. These signals perform as the inputs to the MGW described previously. The MGW-based phase selector is tested on the model network. The tests include a solid and resistive ground fault, and different phase and inter-circuit faults, occurring at different locations and under different loading conditions ($-30° \leq \delta \leq +30°$), respectively. All simulation studies are undertaken for the relay at busbar A.

As explained in Sect. 4.6, fault detection and fault phase selection are related to the setting of threshold ε of the norms, which are calculated from the detail signals of the currents decomposed by MGW. However, the norms should be different according to the loading of the line. Therefore, it is suggested that the setting of threshold ε be proportional to the loading of the line, i.e. to load angle δ. In order to keep reliability under no fault conditions, extensive studies have been conducted to find the optimal setting of threshold ε for different δ. The detailed setting of threshold ε is shown in Table 4.3.

Table 4.3 The setting of thresholds

Load angle: δ	Threshold: ε
$0°$-$\pm 10°$	0.3
$\pm 10°$-$\pm 20°$	0.5
$\pm 20°$-$\pm 30°$	0.6

4.7.1 Single Line Faults

A solid single-line-to-ground fault (SLG) is applied on phase A of circuit I at 100 km from the relay located at busbar A. The fault occurs at 50 ms with a load angle $\delta = -15°$. The six phase currents measured at busbar A and the ground mode current calculated by modal transformation are shown in Figs. 4.17a and 4.18a, respectively. The detail signals of the seven currents are extracted using MGW and their norms are calculated and plotted in Figs. 4.17b and 4.18b, respectively.

The fault phase can be identified by tracking the norm values of the currents. From Figs. 4.17b and 4.18b, it can be seen clearly that only the norm of the detail signal of phase A at circuit I and ground mode G exceeds threshold ε, which indicates a phase-A-ground fault (I AG).

Simulation studies are also carried out under other conditions of SLG faults. Table 4.4 shows the norm of every current, in which the norms higher than threshold ε are highlighted in bold. Therefore, it is easy to determine the fault type. For example, the second row of Table 4.4 shows that the norms of phase B of circuit I and ground G exceed the threshold, respectively. Therefore, it can be concluded that the transmission line suffers from a phase-B-ground fault on the first circuit (I BG).

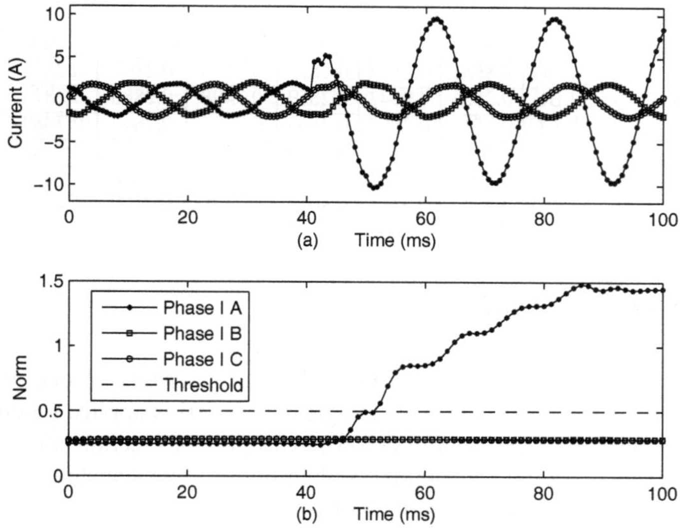

Fig. 4.17 A single line fault on phase A of circuit I with 50 Ω ground-fault resistance at 100 km from busbar A. **a** The fault phase currents of circuit I. **b** The norms of the phase currents

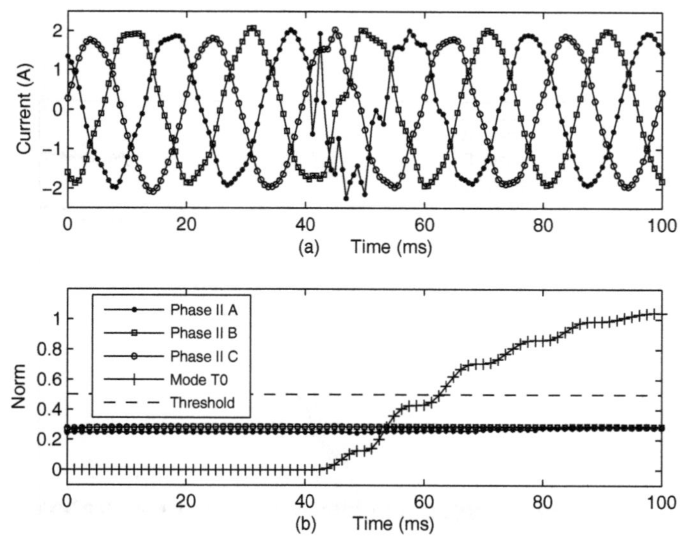

Fig. 4.18 A single line fault on phase A of circuit I with 50 Ω ground-fault resistance at 100 km from busbar A. **a** The fault phase currents of circuit II. **b** The norms of the ground model current and the phase currents

Table 4.4 The norms of currents for different single line faults

Fault condition (R, l, δ)	Fault type	Circuit I			Circuit II			Ground
		E_1 (I A)	E_2 (I B)	E_3 (I C)	E_4 (II A)	E_5 (II B)	E_6 (II C)	E_7 (G)
50Ω, 100 km, −15°	I AG	**0.8532**	0.2872	0.2873	0.2598	0.2867	0.2877	**3.8706**
	I BG	0.2504	**0.9042**	0.2869	0.2498	0.2886	0.2873	**4.5938**
	II CG	0.2511	0.2867	0.2863	0.2505	0.2862	**0.7480**	**4.6570**
100Ω, 50 km, 10°	II AG	0.2116	0.2141	0.1904	**0.5374**	0.1899	0.2085	**3.1696**
	II BG	0.2096	0.2165	0.1888	0.2090	**0.5690**	0.1886	**3.3564**
	I CG	0.2129	0.2106	**0.5015**	0.2128	0.2100	0.1903	**3.5906**

4.7.2 Line-to-Line Faults

The phase selector has also been tested for line-to-line faults on one of the two circuits. A phase-A-C fault with a load angle of $\delta = 20°$ is applied on circuit I at 120 km from the relay at busbar A. As shown in the previous section, after acquiring the detail signals of all the seven currents, their norms are calculated and depicted in Fig. 4.19. Apparently, the norms of phase A and phase C of circuit I increase to a large value after 50 ms and they both exceed the threshold value. This type of fault is referred to as I AC.

Simulation results of several typical faults are illustrated in Table 4.5, from which it can be seen that MGW is applicable to select the fault phases even when the fault occurs in all the three phases.

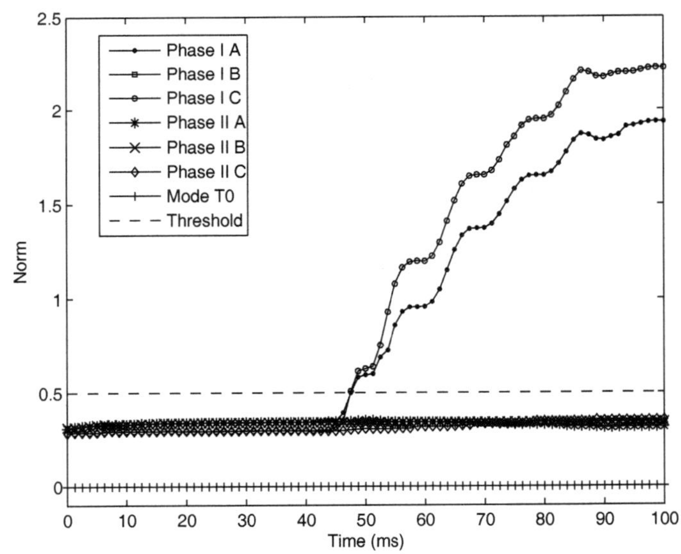

Fig. 4.19 The norms of six phase currents and ground mode current for a phase-A-C fault on circuit I at 120 km from busbar A

Table 4.5 The norms of currents for different line-to-line faults

Fault condition (R, l, δ)	Fault type	Circuit I			Circuit II			Ground
		E_1 (I A)	E_2 (I B)	E_3 (I C)	E_4 (II A)	E_5 (II B)	E_6 (II C)	E_7 (G)
0Ω, 120 km, −20°	I AC	**1.1910**	0.3214	**1.0461**	0.3589	0.3205	0.3524	0.0006
	I BC	0.3570	**1.0159**	**1.2332**	0.3565	0.3227	0.3452	0.0002
	II AB	0.3229	0.3598	0.3451	**1.9105**	**2.2544**	0.3451	0.0094
	I ABC	**1.4117**	**1.3250**	**1.0448**	0.3620	0.3288	0.3477	0.0007
	II ABC	0.3621	0.3288	0.3477	**1.4117**	**1.3251**	**1.0448**	0.0003
50Ω, 80 km, 5°	II AC	0.1678	0.1617	0.1596	**0.8974**	0.1614	**0.9389**	0.0063
	II BC	0.1615	0.1711	0.1450	0.1609	**0.9747**	**0.9941**	0.0011
	I AB	**1.0827**	**1.1237**	0.1421	0.1805	0.1744	0.1419	0.0002
	II ABC	0.1743	0.1747	0.1473	**0.9923**	**1.0695**	**0.8673**	0.0001
	I ABC	**0.9923**	**1.0695**	**0.8673**	0.1743	0.1747	0.1473	0.0001

4.7.3 Simultaneous Faults on Both Line Circuits

4.7.3.1 Cross-Country Faults

The occurrence of a cross-country fault on a double circuit transmission line may initialise serious system instability when phase and ground distance relaying schemes are used for protection, because these schemes assume that the cross-country fault occurs on the two circuits and will cause three-phase tripping for both circuits [28, 40].

In this scenario, a phase-A-ground fault occurs on circuit I and a phase-B-ground fault occurs on circuit II (cross-county fault) at 100 km from the relay at busbar A, and both have a load angle $\delta = -20°$. The norms of the seven detail signals are given in Fig. 4.20, which shows that the norms of phase A of circuit I, phase B of circuit II and ground G rise remarkably when the faults occur. Adjusting the parameters of cross-country fault, the simulation results shown in Table 4.6 demonstrate the adaptability of the developed scheme. Simulation studies are conducted with different fault phases and the results given in Table 4.6 indicate that the MGW-based phase selector is able to distinguish this type of fault and identify the fault phases.

Table 4.6 The norms of currents for different cross-country faults

Fault condition (R, l, δ)	Fault type	Circuit I			Circuit II			Ground
		E_1 (I A)	E_2 (I B)	E_3 (I C)	E_4 (II A)	E_5 (II B)	E_6 (II C)	E_7 (G)
0Ω, 100 km, 20°	I A II BG	**1.2564**	0.3278	0.3451	0.3617	**1.3981**	0.3452	**6.6217**
	I B II CG	0.3571	**1.0673**	0.3458	0.3567	0.3239	**1.0953**	**4.5293**
	I C II AG	0.3604	0.3200	**0.9475**	**1.2325**	0.3200	0.3502	**5.8209**
150Ω, 20 km, 5°	I A II CG	**0.4869**	0.1673	0.1441	0.1677	0.1679	**0.4442**	**2.4752**
	I B II AG	0.1684	**0.5106**	0.1458	**0.5569**	0.1684	0.1463	**2.6413**
	I C II BG	0.1616	0.1677	**0.3933**	0.1641	**0.5766**	0.1431	**2.3642**

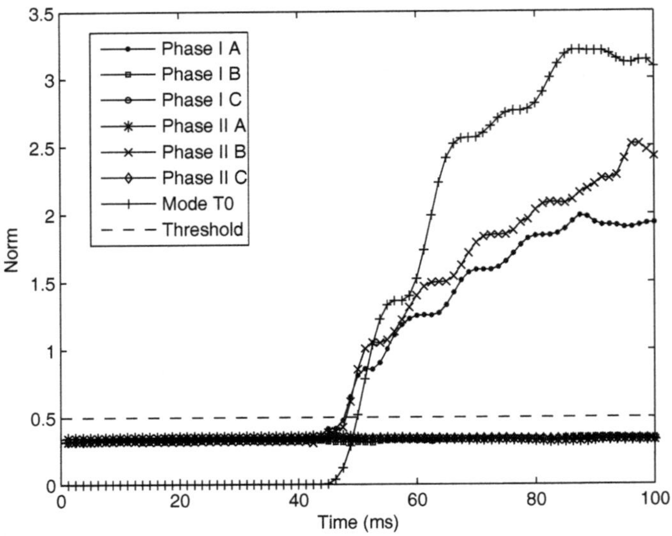

Fig. 4.20 The norms of six phase currents and ground mode current for a cross-country fault on phase A of circuit I and phase B of circuit II at 100 km from busbar A

4.7.3.2 Same-Phase Faults

Same-phase faults are a special case where a line-to-ground fault occurs on one phase of circuit I and, at the same time, on the same phase of circuit II. As an example, a phase-A-ground fault on circuit I and a phase-A-ground fault on circuit II are simulated at 150 km from busbar A with a load angle of $\delta = -30°$. From the norms for the fault currents shown in Fig. 4.21, it can be seen explicitly that the norms of phase A of circuit I and phase A of circuit II are identical and they exceed the threshold value after the fault occurs, and so does the norm of the ground mode current. Simulation results of other same-phase faults are given in Table 4.7.

Table 4.7 The norms of currents for different same-phase faults

Fault condition	Fault	Circuit I			Circuit II			Ground
(R, l, δ)	type	E_1 (I A)	E_2 (I B)	E_3 (I C)	E_4 (II A)	E_5 (II B)	E_6 (II C)	E_7 (G)
0Ω, 150 km, −30°	I A II AG	**0.7040**	0.5036	0.5060	**0.7040**	0.5041	0.5060	**6.7894**
	I B II BG	0.4363	**0.7287**	0.5097	0.4363	**0.7288**	0.5102	**9.0432**
	I C II CG	0.4430	0.5041	**0.6688**	0.4430	0.5045	**0.6688**	**9.8792**
50Ω, 50 km, 10°	I A II AG	**0.6015**	0.2180	0.1884	**0.6015**	0.2178	0.1884	**6.6292**
	I B II BG	0.2095	**0.6687**	0.1938	0.2095	**0.6687**	0.1940	**7.4032**
	I C II CG	0.2167	0.2102	**0.4986**	0.2167	0.2096	**0.4986**	**8.1863**

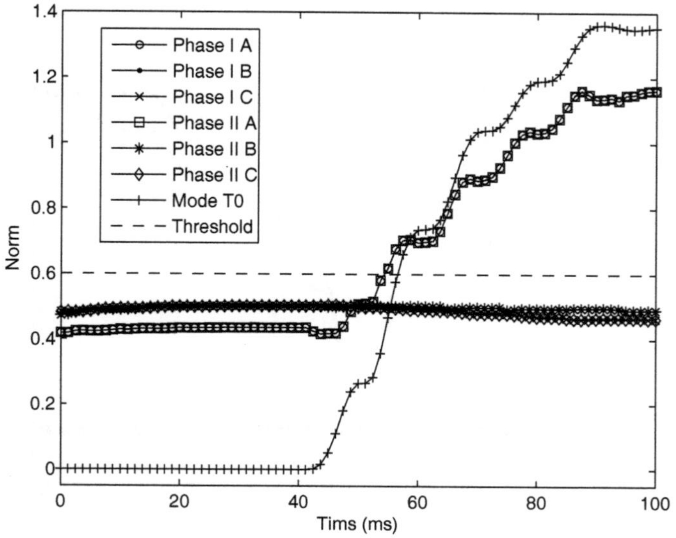

Fig. 4.21 The norms of six phase currents and ground mode current for a phase-A-ground fault on circuit I and a phase-A-ground fault on circuit II at 150 km from busbar A

4.7.3.3 Clashing-Conductor Faults

Phase-to-phase faults on both circuits are simulated to test the MGW-based phase selector as well. This type of fault is called a clashing-conductor fault. A phase-A-of-circuit-I-to-phase-B-of-circuit-II fault at 50 km from the relay located at busbar A, with a load angle of $\delta = 10^\circ$, is simulated as a clashing-conductor fault. Figure 4.22 plots the fault phase selection process and the results show that the phase selector detects the fault phases accurately. Likewise, several other clashing-conductor faults are also simulated to examine the robustness of the phase selector and the results are listed in Table 4.8.

4.8 Summary

This chapter presented two examples of the application of MM-based methods for the protection of transmission lines. In the first example, the ADRA was introduced, which contains three tasks: fault detection, classification and location. The detailed procedure of performing the three tasks has been discussed. An MFD has been designed to extract the feature of transient current waveform so as to detect the fault occurrence. It is able to provide fast calculation and accurate detection results. The fault classification is achieved by comparing the different Euclidean norms of the MFD outputs of different types of faults, which is a simple yet reliable strategy. The

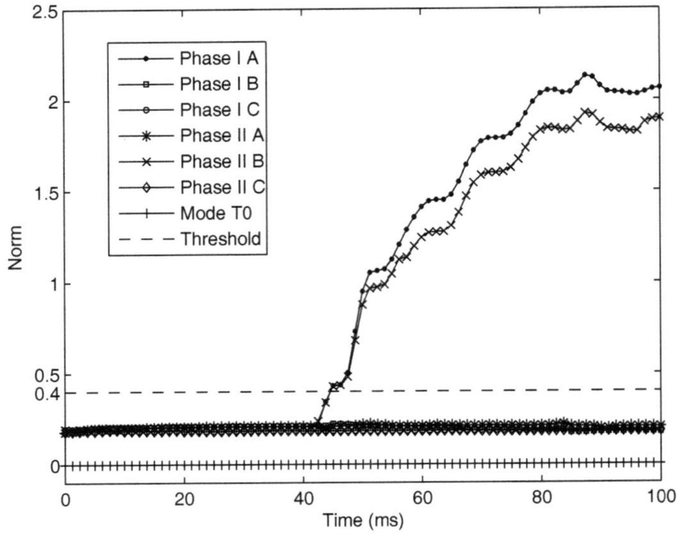

Fig. 4.22 The norms of six phase currents and ground mode current for a clashing-conductor fault on phase A of circuit I and phase B of circuit II at 50 km from busbar A

Table 4.8 The norms of currents for different clashing-conductor faults

Fault condition	Fault	Circuit I			Circuit II			Ground
(R, l, δ)	type	E_1 (I A)	E_2 (I B)	E_3 (I C)	E_4 (II A)	E_5 (II B)	E_6 (II C)	E_7 (G)
0Ω, 50 km, $10°$	I A II B	**1.3929**	0.2241	0.1848	0.2249	**1.2180**	0.1855	0.0001
	I B II C	0.2099	**1.1257**	0.1961	0.2106	0.2156	**1.0953**	0.0016
	I BC II B	0.2103	**0.7619**	**1.4105**	0.2099	**0.6485**	0.1987	0.0006
	I BC II C	0.2099	**1.4776**	**0.7461**	0.2103	0.2170	**0.6165**	0.0039
	I A II BC	**1.3581**	0.2272	0.2030	0.2210	**1.3722**	**1.1533**	0.0002
100Ω, 120 km, $-20°$	I AB II BC	**1.3514**	**0.8621**	0.3457	0.3597	**0.7925**	**1.1698**	0.0002
	I ABC II A	**0.8408**	**1.2081**	**1.3031**	**0.7579**	0.3516	0.3465	0.0002
	I BC II BC	0.3565	**0.7567**	**0.9478**	0.3565	**0.7567**	**0.9478**	0.0002
	I ABC II BC	**1.3283**	**0.9178**	**0.7665**	0.3597	**0.8411**	**0.7291**	0.0002
	I ABC II ABC	**0.8474**	**0.8765**	**0.7596**	**0.8473**	**0.8765**	**0.7596**	0.0002

fault location is calculated within each sampling interval, based on the instantaneous circuit model of the transmission system, which can be implemented in real time as a transient protective relaying approach.

The identification of a variable tripping zone has been discussed to improve both the speed and reliability of distance relaying. Working as an adaptive mechanism, ADRA requires only a few samples in the early stage of recursive calculation, which ensures a fast response to faults occurring at a location close to the relay. To deal with the faults occurring in longer distance, more samples have been involved to estimate the fault location. Compared with conventional distance relaying methods, ADRA covers a larger protection zone due to its adaptive feature. This feature is also attributed to its ability in estimating the fault location without knowing the ground-fault resistance.

The performance of ADRA has been comprehensively tested for a number of different types of faults, which are simulated with consideration of various of fault distance, load angle and ground-fault resistance, respectively. The simulation results have shown that different fault distances simulated both inside and outside the first protection zone of the relay can be correctly identified by ADRA. The fast tripping time of ADRA for typical fault scenarios has been achieved, and the results have indicated that ADRA is able to take a tripping action in less time than that required by conventional distance relaying methods. Moreover, a wider zone of the transmission line can be reliably protected by ADRA.

The second example was the design of a phase selector for fault classification of a double circuit transmission line. It demonstrates that the MGW-based phase selector is able to identify the fault phases of the double circuit transmission line in spite of the presence of mutual coupling between the two circuits. In order to construct the phase selector, the phase currents of the double circuit are firstly decoupled by means of a model transformation. The ground mode current and the six phase currents are then processed by MGW which is able to extract the features of transient fault currents and obtain the detail signals for fault phase selection following a criterion given. The phase selector has been successfully tested for all types of faults in simulation studies. The simulation results validate that the phase selector is robust to the variation of system parameters, such as fault location, ground-fault resistance, and load angle changes.

As a fast calculation tool, MGW can effectively and rapidly extract the features of transient currents. The simple computation required by MGW makes it convenient to implement in real time. The application of MGW to phase selection reveals that it has great potential for feature detection in other cases of power system operation and control.

Chapter 5
Transformer Protection

Abstract This chapter presents two MM-based schemes for the identification of transformer magnetising inrush. The first scheme detects the inrush current using a morphological decomposition scheme (MDS) to extract the singularity peculiar to the asymmetric inrush waveform, while attenuating other irrelevant features [135]. By quantifying the extracted features, the inrush current is then identified using a set of simple criteria. The second is a multi-resolution decomposition scheme (MRDS), which is developed based on MDS but performs faster than MDS [93]. It decomposes a current signal into multi-resolution levels using synthesis and analysis operators of MM. MRDS is able to discriminate between inrush and internal fault currents even in the case of an inrush that contains a low second harmonic component and an internal fault current that has a high second harmonic component. Simulation studies have been undertaken to evaluate the two schemes based on a three-phase high voltage power transformer operating under different conditions. The test results show that the schemes are able to identify inrush currents reliably and provide an alternative solution for transformer differential protection.

5.1 Introduction

Power transformers are a class of very expensive and important apparatus in electric power systems [79]. The design of relays for power transformer protection has been a challenging problem for many years. Traditionally, a differential protection relay is always selected as a primary protection strategy for most power transformers [41, 53]. In a differential protection relay, differential currents can provide significant information about the feature of the current flowing through power transformer windings. Such information may include frequency contents, transient features, periodicity, non-periodicity and frequency-time relations. There are many factors that should be considered when attempting to design a differential protection relay for power transformers, such as the magnetising inrush current, overexcitation and CT saturation.

The main challenge of power transformer protection is to avoid a false tripping action in the case of the occurrence of a magnetising inrush current, i.e. the ability of the protective relaying algorithm to distinguish different magnetising inrush currents from internal faults, external faults and normal currents [61]. Inrush current is frequently encountered when the transformer is energised. While a transformer is connected to a power source, the inrush current may reach 10 to 20 times of the rated value of transformer current, which is as high as that of internal fault currents. Therefore, it is difficult to distinguish between them, and a malfunctioned tripping action may be taken by conventional differential relays in this case. Such a malfunction will affect both the reliability and stability of the whole power system.

The conventional methods developed for detection of inrush currents are based on identifying the second harmonic component and are commonly used for blocking differential relays, since the inrush currents normally contain a large second harmonic [138, 142]. However, the second harmonic of an inrush current is significantly decreased when the core steel of modern power transformers has been improved [129]. Moreover, the second harmonic component may also be produced by internal faults of the power transformer due to the presence of a shunt capacitor or the distributive capacitance on a long EHV transmission line connected to the transformer [89]. Consequently, the techniques based on second harmonic identifications will not be sufficiently effective when they are used in differential protection relays, as they cannot accurately identify transformer inrush, using the content of the second harmonic component. Furthermore, second harmonic-based methods slow down the operating speed of the transformer relays, since a wide sample window is required for calculating the second harmonic component using an FT technique.

The methods based on the direct detection of the distortion of waveform shapes have been investigated to find the difference between the internal fault and inrush currents. A major group of these methods is based on the identification of the peaks of differential currents. If successive peaks of the differential current fail to occur over a period of 7.5–10 ms, the relay operation will be blocked [114]. Recognising the length of the time interval between the zero-crossing points of the differential current has been considered as another well-known principle for distinguishing between the internal fault and inrush currents [8, 36]. For inrush currents, the transformer relay will be blocked if the time interval, during which the amplitude of the current is close to zero, is longer than a threshold. However, the interval may decrease due to CT saturation. Recently, a blocking scheme called waveform correlation was introduced in [2]. It compares the symmetry between the first half of a cycle and the second half of the cycle of the waveform using a complete fundamental cycle of sampling data. In this method, internal fault and inrush currents can be distinguished if the correlation coefficient between the two groups of data obtained from the first half of the cycle and second half of the cycle, respectively, exceeds a given threshold. However, using this method, it is still difficult to identify the inrush currents that are symmetric. Moreover, a full cycle of data is required for correlation calculation. Recently, several alternative algorithms were developed for the detection of inrush currents. In [102], [107] and [110], an ANN was trained to distinguish between inrush and internal fault currents. A fuzzy logic method was suggested to develop differential protection relays [143]. For the same purpose, a method of frequency transient detection was investigated [21]. Moreover, some algorithms based on WT were reported in [54] and [149]. However, the wavelet method involves complex calculation.

Most of these approaches are still liable to cause the malfunction of the relay in the cases where the transformer is energised under complex connection conditions, in particular in the case where the inrush current has a low second harmonic component and the internal fault current has a high second harmonic component.

This chapter describes two MM-based schemes to identify the inrush current. The schemes are fundamentally different from conventional methods, as they decompose the signal based on time-domain features instead of in the frequency domain. Since they work directly upon the geometric characteristics of the input signal, there is no need for transform techniques such as FT to process the input signal. Before comprehensively discussing the MM-based schemes, the transformer differential protection and the characteristics of magnetising inrush currents are introduced in the following sections.

5.2 Transformer Differential Protection

Large transformers are generally protected by percentage current differential relays with restraining algorithms based on second and sometimes fifth harmonics [144, 151]. The percentage current differential protection has been recognised as the most principal theory of transformer protection and has had a proven record of reliable operation for many years. A percentage differential function is applied to the fundamental component of the currents to decide whether an internal fault has occurred. It converts the primary and secondary currents to a common base and compares the operating current with a restraining current. The difference between the operating and restraining currents is small for normal operating conditions and external faults, while it becomes significant during the internal fault.

The operating current of percentage current differential protection can be obtained by:

$$I_{\mathrm{op}} = |I_{\mathrm{m}} + I_{\mathrm{n}}|, \tag{5.1}$$

where I_{m}, I_{n} are the primary current and the secondary current, respectively, and I_{op} is proportional to the fault current for internal faults and approaches zero for other operating (ideal) conditions.

Different alternatives have been applied to obtain the restraining current, I_{rt}. The most common modes are:

$$I_{\mathrm{rt}} = \frac{1}{2}|I_{\mathrm{m}} - I_{\mathrm{n}}|, \tag{5.2}$$

$$I_{\mathrm{rt}} = \frac{1}{2}(|I_{\mathrm{m}}| + |I_{\mathrm{n}}|), \tag{5.3}$$

$$I_{\mathrm{rt}} = \max(|I_{\mathrm{m}}|, |I_{\mathrm{n}}|). \tag{5.4}$$

The percentage differential protection scheme can be denoted as:

$$\begin{aligned} &I_{op} > I_{op0} && I_{rt} < I_{rt0}, \\ &I_{op} > K_1(I_{rt} - I_{rt0}) + I_{op0} && I_{rt} > I_{rt0}, \end{aligned} \tag{5.5}$$

where K_1 is a proportional constant. Normally, $K_1 = (0.4 \sim 0.7)$, I_{rt0}, I_{op0} are setting values, and $I_{rt0} = (0.2 \sim 0.6)I_{rated}$, $I_{op0} = (0.8 \sim 1.0)I_{rated}$, where I_{rated} is a rated current of the transformer.

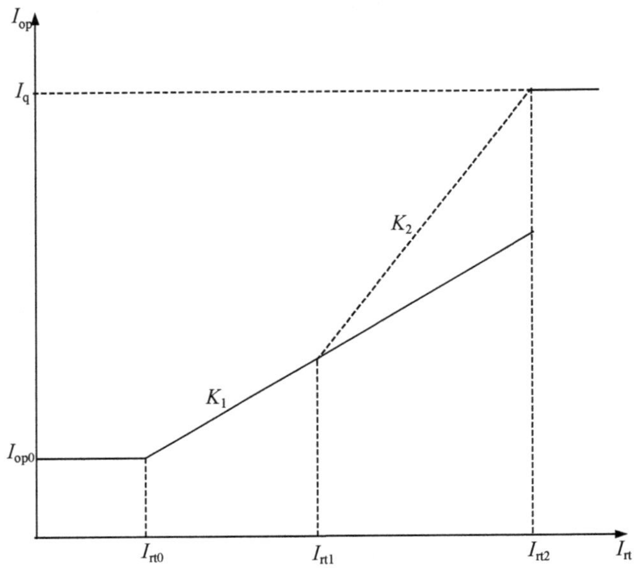

Fig. 5.1 The characteristics of percentage restraint of differential protection

The characteristics of the differential protection scheme are plotted in Fig. 5.1, where there are two straight lines given with a slope of K_1 and a slope of K_2, which range from I_{rt0} to I_{rt1} and from I_{rt1} to I_{rt2}, respectively, and a horizontal straight line defining the relay minimum pickup current, I_{op0}. The relay operating region is located above the slope, and the restraining region is below the slope.

A dual-slope percentage characteristic provides further security for external faults with CT saturation. It is represented as a dashed line in Fig. 5.1. The slope of the dashed line is denoted as K_2 and it normally takes a value between 0.5 and 0.75. The percentage differential protection scheme can be re-written as:

$$\begin{aligned} &I_{op} > I_{op0} && \text{if } I_{rt} < I_{rt0}, \\ &I_{op} > K_1(I_{rt} - I_{rt0}) + I_{op0} && \text{if } I_{rt0} \le I_{rt} \le I_{rt1}, \\ &I_{op} > K_1(I_{rt1} - I_{rt0}) + K_2(I_{rt} - I_{rt1}) + I_{op0} && \text{if } I_{rt} > I_{rt1}. \end{aligned} \tag{5.6}$$

The dual-slope percentage pattern adds a restraint area and avoids mal-operation caused by CT saturation. In comparison with a single-slope percentage scheme, the dual-slope percentage current differential protection can be regarded as a better curve fitting of transformer operational principles.

5.3 Transformer Magnetising Inrush

Magnetising inrush currents in power transformers result from an abrupt change of the magnetising voltage. Although usually considered as the result of energising a transformer, the magnetising inrush may also be caused by [81, 85]:

1. the occurrence of an external fault,
2. voltage recovery after clearing an external fault,
3. changes of the character of an external fault, and
4. out-of-phase synchronising of a nearby generator.

The magnetising current disturbs the balance between the currents at the transformer terminals, and is therefore identified by the differential relay as a 'false' differential current. The relay, however, must remain stable during inrush conditions. In addition, considering the lifetime of the transformer, tripping out during an inrush condition is a very undesirable situation (breaking the current of a pure inductive loading generates high over-voltage, which may jeopardise the insulation of the transformer and be the indirect cause of an internal fault).

The magnetising inrush current cannot be easily detected, thus the characteristics of this type of current are very complex and difficult to analyse. It depends on many factors such as the instant of connecting the supply to the power transformer, the residual flux, the type of winding connection, the core type material and the type of relation between current and voltage.

5.3.1 The Magnitude of Magnetising Inrush Current

When a transformer is initially energised, there is a transient inrush current establishing the magnetic field of the transformer. Considering a transformer that is energised from a sinusoidal voltage, the steady-state flux Φ is the integral of voltage shown as follows:

$$\Phi = \frac{1}{N}\int \sin\omega t \mathrm{d}t = \frac{-1}{\omega N}\cos\omega t, \tag{5.7}$$

where N is the number of turns of the winding and ω is the angular frequency of the voltage signal. The instantaneous magnitude of the core flux Φ at the instant of energising is defined as the residual flux Φ_{r}. The amount of offset of the sinusoidal flux generated by the voltage is dependent upon the point of the voltage wave where the transformer is energised. Assuming that the peak magnitude of the normal core flux is labelled as Φ_m, the peak core flux therefore reaches $2\Phi_m + \Phi_r$, which will drive the core into saturation. The worst case is that the transformer is energised at the zero-crossing point of the voltage wave with a residual flux of Φ_m. In this case, the saturation is even greater and a significant inrush current is then excited.

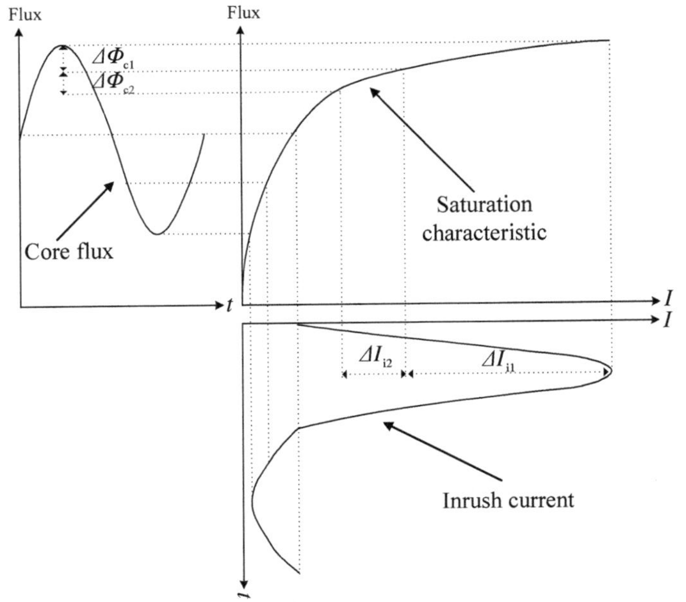

Fig. 5.2 Derivation of the inrush current waveform from the saturation curve

The formation of inrush is illustrated in Fig. 5.2. Each point along the core flux waveform (top left graph, Fig. 5.2) has a corresponding point on the saturation curve (top right graph, Fig. 5.2), which is plotted against the inrush current (bottom graph, Fig. 5.2) in time evolution. The magnetic core flux is mapped to the inrush current via the saturation curve. It can be clearly seen that the inrush current is not a sinusoidal waveform, because the saturation curve is not linear. The first half of a cycle of the inrush current has a peak occurring at the maximal flux, which is not symmetrical to the second half of the cycle. The inrush current in the first half of the cycle is mostly generated by the flux-current trajectory above the knee point of the saturation curve, while most of the second half of the cycle of the current is mapped from the trajectory under the knee point. In contrast to a sinusoid, the inrush current is distorted because of magnetic saturation. Moreover, as shown in Fig. 5.2, two equal increments of core flux in a sinusoidal wave, $\Delta\Phi_{c1}$ and $\Delta\Phi_{c2}$, generate increments of dramatically different height in inrush current, as ΔI_{i1} and ΔI_{i2}, respectively. In fault currents, the two increments have the same height due to linear mapping. That is the major cause of difference between inrush and fault currents, in particular in the features of their changes in slope and symmetries. Notwithstanding, these features can be extracted by MM-based schemes.

A typical inrush current waveform is given in Fig. 5.3, where a large and long-lasting DC component can be observed, which has large peak values at the beginning and decays substantially after a few tenths of a second, but only decays fully after several seconds. The shape, magnitude and duration of the inrush current depend on several factors, including:

1. the size of the transformer,
2. the impedance of the energising system,
3. the magnetic properties and remanence of the core, and
4. the phase angle when the transformer is switched on.

A detailed discussion about how these factors affect the inrush waveform can be found in [81].

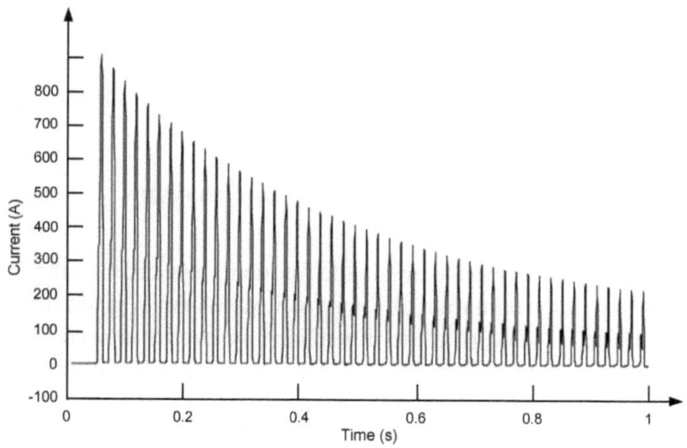

Fig. 5.3 A typical inrush current waveform

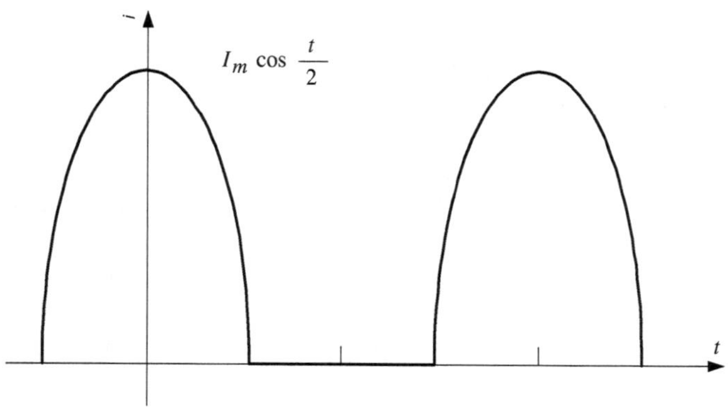

Fig. 5.4 An idealised inrush current waveform for spectral analysis

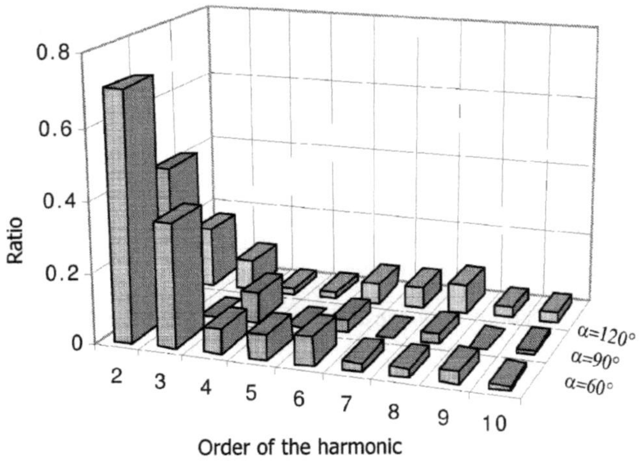

Fig. 5.5 The harmonic content of the idealised inrush current with $\alpha = 60°$, $90°$ and $120°$

5.3.2 Harmonics of Magnetising Inrush Current

For a single-phase transformer, to obtain the frequency spectrum of the inrush waveform, an analytical approximation of the inrush waveform is given in Fig. 5.4. The waveform between $-\alpha$ and α is the inrush current due to saturation of the transformer air core, between α and $2\pi - \alpha$ the intermittent angle of the waveform, known as the gap. The angle α is used to facilitate the modelling of an actual inrush current.

The amplitude of the nth harmonic, A_n, of the waveform shown in Fig. 5.4 can be calculated as

$$A_n = \frac{I_m}{\pi}\left[\frac{1}{n+1}\sin((n+1)\alpha) + \frac{1}{n-1}\sin((n-1)\alpha) - 2\cos\left(\frac{\alpha}{n}\right)\sin(n\alpha)\right], \quad (5.8)$$

where I_m is the peak value of the inrush current [81].

Figure 5.5 presents the spectrum of the signal, shown in Fig. 5.4, calculated by (5.8) with $\alpha = 60°$, $90°$ and $120°$, respectively. It is clear that the second harmonic always dominates because of the large DC offset. However, the amount of the second harmonic may drop below 20%. The minimal content of the second harmonic depends mainly on the knee point of the magnetising characteristic of the core; the lower the saturation flux density, the higher the amount of the second harmonic. Modern transformers built with improved magnetic materials have high knee points, and therefore, their inrush currents display a relatively low amount of the second harmonic. Since the second harmonic is the basic restraining criterion for stabilising differential relays during inrush conditions, certain difficulties may arise when protecting these transformers [61].

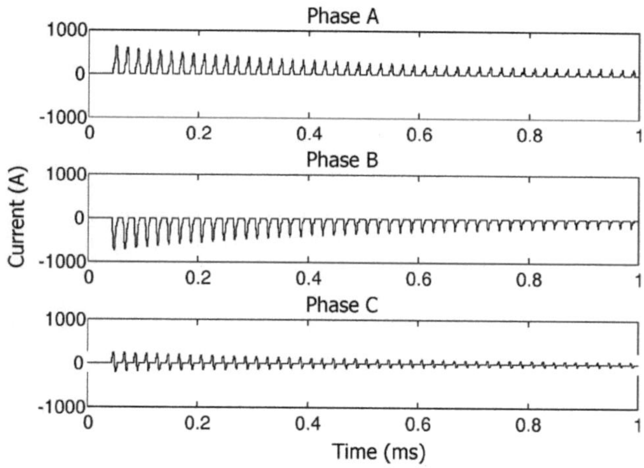

Fig. 5.6 The inrush currents in a three-phase transformer

Inrush currents measured from a three-phase transformer may differ considerably from the above single-phase transformer, as:

1. the angles of the energising voltages are different in different phases;
2. when a delta-connected winding is switched in, the line voltages are applied as the magnetising voltages;
3. in the above case the line current in a given phase is the vector sum of the other two windings' currents; and
4. depending on the core type and other conditions, it is possible that only some of the core legs are saturated.

Therefore, for the current in a given phase or a grounded neutral point of the transformer, it is either similar to the single-phase inrush pattern, as shown in Fig. 5.3, or becomes a distorted but oscillatory waveform. In the latter case, the amount of the second harmonic may drop dramatically, which causes problems for differential relaying. In Fig. 5.6, which illustrates the waveforms of energising a three-phase transformer, the currents in phases A and B exhibit a typical rectified inrush pattern, whereas phase C is an oscillatory waveform.

Due to the exponentially decaying DC offset in great magnitude, the inrush current waveform is likely to be distorted by the saturation of CT. In such a condition, the CT's secondary current may exhibit a certain level of distortion as shown in Fig. 5.7, with a considerably reduced amount of the second harmonic [82], and in the most adverse conditions, the gap disappears.

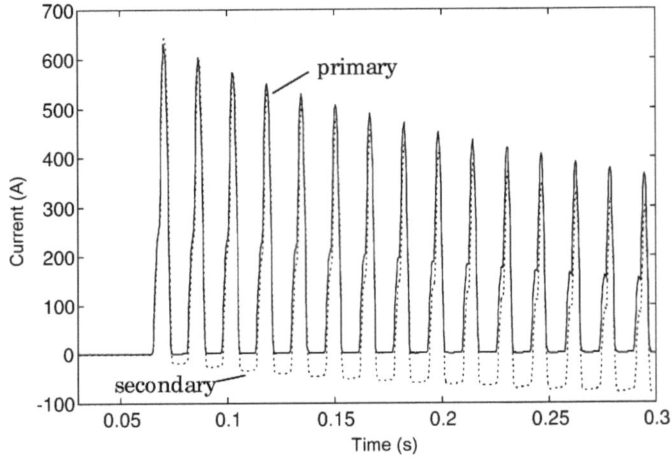

Fig. 5.7 The primary and secondary currents of a saturated CT

5.3.3 The Second-Harmonic Restrained Differential Protection of Transformers

Modern means of restraining differential relays during magnetising inrush conditions recognise the inrush pattern either indirectly (harmonic analysis) or directly (waveform analysis) [80].

The harmonic restraint method is the classical way to restrain the relay from tripping during magnetising inrush conditions. As mentioned above, the inrush current is rich in harmonics, and generally, low levels of harmonics enable tripping, while high levels indicate inrush and restrain the relay. For digital relays, this may be written as

$$T = \text{true}, \ \text{if } I_{\text{CH}} < \delta I_{\text{CD}}, \tag{5.9}$$

where T denotes tripping permission from the inrush detector, I_{CH} is the combined harmonic component in the differential current, and I_{CD} denotes the combined differential current and δ is a threshold. The condition given in (5.9) originates a whole family of algorithms using a variety of approaches for combining currents I_{CH} and I_{CD}.

In the simplest approach, the amplitude of the second harmonic is considered as the combined harmonic signal, while the amplitude of the fundamental frequency component in the same phase is used as the combined differential current.

Harmonic restraint in general, regardless of the method of composing the combined harmonic and differential signals, displays certain limitations. Firstly, the estimator of the harmonic component (usually the second harmonic only) needs a certain amount of time for accurate estimation of the amplitude. Even if the harmonic is not present in the differential signal at all, the ratio of I_{CH} to I_{CD} in (5.9) is initially significantly overestimated (until the fault data fills out the estimator data window). This means that the harmonic restraint usually will not permit tripping for a time approximately equal to the data window length of the estimators (typically one cycle). Secondly, as mentioned above, in modern transformers the proportion of higher harmonics in the magnetising current may drop well below 10% (the second harmonic can be as low as 7%, while the total harmonic content is at a level around 7.5% [18, 79]). Under such circumstances, the setting δ in (5.9) should be adjusted to be under 7%. This, however, may lead to delayed or even missing operations of the relay, due to the harmonics in the differential currents during internal faults accompanied by saturation of the CTs. Thirdly, the second harmonic-to-fundamental ratio may temporarily (for several cycles) drop below the threshold of 20% used in most relays [53].

For waveform-based restraining methods, the most commonly used technique is to detect the periods of low and flat values, i.e. the gap, in the waveform. In this method, the hypothesis of magnetising inrush may be ruled out if the differential current does not include in its every cycle a period that lasts no less then one quarter of a cycle while its shape is both flat and close to zero. This relaying principle was known in the era of static relays and certain analogue schemes have been developed for implementing it [41, 53].

This form of direct waveform restraining regardless of its implementation has weaknesses [81]. Firstly, the recognition of an internal fault versus magnetising inrush takes one full cycle. Secondly, the CTs, when saturated during inrush conditions, change the shape of the waveform within the gap period as shown in Fig. 5.7 and may cause a false tripping. Thirdly, during severe internal faults, when the CTs are saturated, their secondary currents may also show periods of low and flat values, causing the relay to malfunction.

5.4 Morphological Identification of Inrush

5.4.1 A Morphological Decomposition Scheme for Inrush Detection

5.4.1.1 Extraction of the Asymmetric Features

To extract the asymmetric features associated with the inrush current, an MDS suggested in [34, 111] is employed to decompose the waveform into various levels of detail. Let f be a signal; to decompose f as $f = \sum_i f_i$ with a set of components $\{f_i, i = 1, 2, \ldots\}$, MDS can be applied for obtaining a family of residues of f. If the residue of two transforms, ϕ and ψ, is defined as $R_{\{\phi,\psi\}}(f) = \phi(f) - \psi(f)$, with an anti-extensive and positively defined transform γ, the general representation of MDS can be defined as:

1. The family of residues of f with respect to γ is $\{r_i, i = 1, 2, \ldots\}$, where $r_1 = f$ and $r_{i+1} = R_{\{r_i,\gamma(r_i)\}}$ until $\gamma(r_i)$ is a null signal.

2. The set $\{f_i\}, f_i = \gamma(r_i)$ is the residual representation of f with respect to the constructive transform γ. Reconstruction f^* of f can be obtained as $f^* = \Sigma_i f_i$. The partial j-components' reconstruction is given by $f_j^* = \Sigma_{i=1}^{j} f_i$.

The family of residues can be considered as the result of the identity transform $\mathrm{id}(r) = r$ and the constructive transform γ. When the constructive transform is recursively applied to the signal, a residue family can be uniquely determined. With j partial reconstruction, f can be represented in arbitrary levels of detail. The reconstruction is non-decreasing by definition, when the final residue r_{j+1} is null, a complete reconstruction f_j^* can be obtained since $f = f_j^* + r_{j+1}$. A non-converging reconstruction can appear if r_{j+1} is not null and the transform $\gamma(r_j)$ is a null function. Its error has to be determined.

There are a large number of transforms fulfilling the above constraints. For our studies of inrush current identification, one of them should be chosen such that the corresponding residual representation is optimal, when the signal is partially reconstructed, in obtaining the singularities peculiar to the inrush signal.

Morphological erosions and openings are anti-extensive transforms if their SEs satisfy $g(0) > 0$. Furthermore, they are constructive transforms if the region of support of the erosion $[f \ominus kg](x)$ is limited to those points x of the domain for which $[f \ominus kg](x) \geq 0$ [111]. Let r_{kg} be the family of openings of r with kg, the homothetics of an SE, g. As k increases, the openings are a more simplified version of the original function and the details that are less than kg are discarded. This reason suggests selection of the maximal scale opening as the constructive transform. If κ is the corresponding maximal value of k, the constructive transform is $\gamma(r) = \max_k(r_{kg}) = r_{\kappa g}$. Following the approach mentioned above, the definition of MDS is recovered [111]. The family of residues is $r_{i+1} = R_{\{r_i, r_{\kappa_i g}\}}$ and the components of the representation are

$$f_i = \gamma(r_i) = (r \ominus \kappa_i g) \oplus \kappa_i g. \tag{5.10}$$

Information on the shape of the signal is captured by g while κ_i provides information on the size. The components are maximal functions in the sense that they describe the largest scaled version of the SE, g, that can still be fitted inside the current residue. The maximal SE is an invariant for the residue. The support of the components f_i are functions $l_i, l_i = r_i \ominus \kappa_i g$. By using l_i functions, one can obtain $f_i = l_i \oplus \kappa_i g$. The function l_i represents the loci of the centres of the maximal SE in the current residue and l_i is positive on its domain. The reconstructed f is given by

$$f^* = \Sigma_i f_i = \Sigma_i (r_i \ominus \kappa_i g) \oplus \kappa_i g = \Sigma_i l_i \oplus \kappa_i g. \tag{5.11}$$

The morphological representation is translation invariant and, in the continuous setting, scale invariant and rotation invariant (if the SE is rotation invariant). The quality of the reconstruction depends on the shape of the SE and its size. The details that are smaller than the SE are lost and the reconstructed function is a smoothed version of the original one.

Consider a typical cycle of the inrush current, as illustrated in Fig. 5.8, where the 50 Hz current signal I is normalised as I/I_{n}. I_{n} is the rated secondary current of the transformer. It is clear that the asymmetries of the waveform are exhibited in both the amplitude and the duration of the positive and negative segments. If the signal is decomposed to have the peak around A_{max} and the valley around A_{min} separated from other components, both with a flat base of given length, then the difference in the amplitude of A_{max} and A_{min}, as well as the interval between them, can be quantified and compared with those in the internal fault current and normal power system operation conditions.

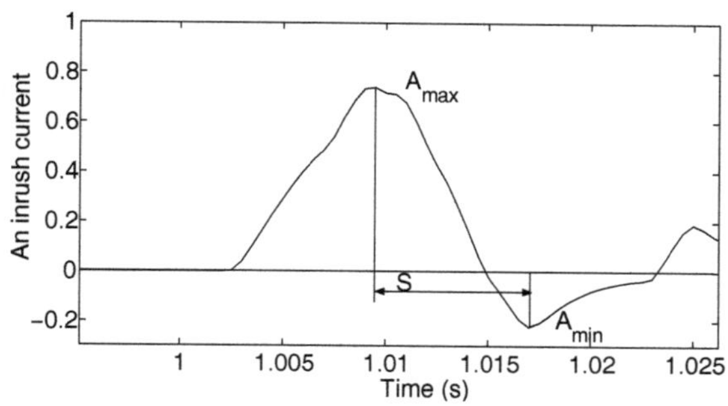

Fig. 5.8 A typical inrush current waveform

Compared with the inrush current, the waveform of the internal fault current, as shown in Fig. 5.9, is a regular periodic signal with an exponentially decaying DC offset. If the effect of DC offset is removed, the waveform becomes a normal sinusoid. When the same decomposition as that mentioned above is applied, the extracted peaks and valleys should have similar amplitudes, with intervals approximately equal to half of the power system cycle.

To perform such a decomposition for extracting the peaks and valleys of the transformer current signals, MDS is applied as illustrated in Fig. 5.10.

The input current waveform is first transformed to obtain two signals f and f', defined as

$$\begin{aligned} f &= I + i_0 \\ f' &= -I + i_0 , \end{aligned} \tag{5.12}$$

where i_0 is a preset constant. i_0 is necessary since the morphological signal decomposition requires that the input signal f satisfies $[f \ominus kg](t) \geq 0$ [111]. The value of i_0 is not crucial as long as it results in both f and f' being positive. The current signal contains both peaks and valleys, but the erosion operation in the decomposition procedure can only extract the peaks of a signal. Therefore, the inverted input signal, f', is decomposed separately and then inverted again to obtain the valleys of the signal.

The SE, g, used for decomposition is a simple zero-valued flat line with a length of 0.02 s, with its origin at the centre. However, the maximal number of iterations of decomposition, κ, does not need to be large enough to obtain a full reconstruction of the original signal, as long as it can extract the peaks of the waveform for measure-

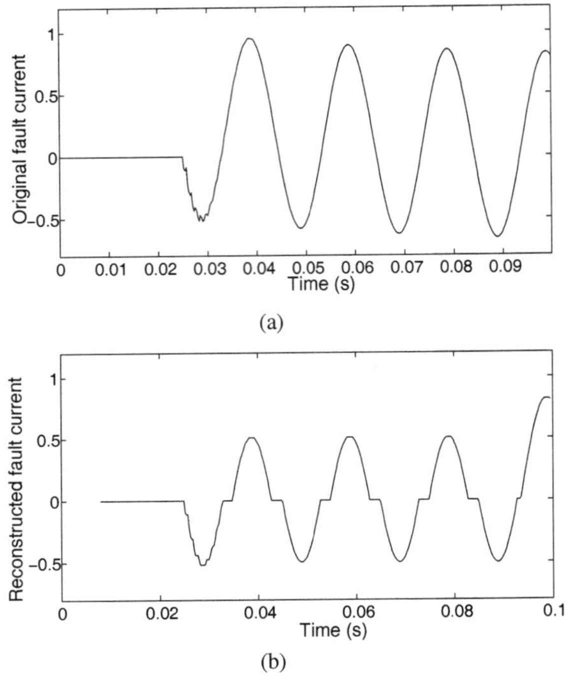

Fig. 5.9 The internal fault current and its reconstruction. **a** Internal fault current waveform. **b** Partial reconstruction of internal fault current with $\kappa = 0.08$

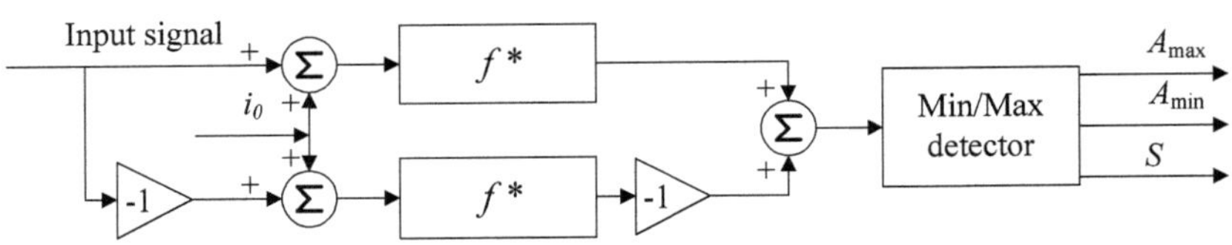

Fig. 5.10 Block diagram of the developed method

ment. In our study, we choose 0.08 s as the final SE; the reason will be discussed in following section. Therefore, there are three iterations of MDS, with SE length at 0.02, 0.06 and 0.08, respectively.

5.4.1.2 Classification and Decision

When a signal is decomposed to the given level, simple criteria can be employed to quantify the asymmetry in the signal, as follows:

$$\delta_I = \frac{|A_{\max} - |A_{\min}||}{A_{\max} + |A_{\min}|} \times 100\%, \tag{5.13}$$

$$\delta_t = \frac{|S - 0.01|}{0.01} \times 100\%, \tag{5.14}$$

where δ_I represents the amplitude asymmetry between the extracted peaks $A_{\max}$ and valleys $A_{\min}$, whereas δ_t is the peak-to-valley interval asymmetry normalised by half of the power system cycle, which is 0.01 s in the case of 50 Hz. The higher δ_I and δ_t, the stronger the asymmetry in the signal. Therefore, if both δ_I and δ_t exceed a pre-defined threshold, an inrush current can be identified. Based on our practical experiment, we use 10% and 20% as the threshold for δ_I and δ_t, respectively.

5.4.2 A Multi-resolution Decomposition Scheme for Inrush Identification

5.4.2.1 Introduction of MRDS

MRDS was introduced in Sect. 2.4.1 based on the concept of the morphological wavelet. It is used here to identify inrush currents. The scheme is able to decompose differential currents into a series of components for the purpose of inrush identification.

In MRDS, the analysis operators, $\psi_j^{\uparrow}$ and $\omega_j^{\uparrow}$, and the synthesis operator $\Psi_j^{\downarrow}$ are defined as:

$$\begin{aligned} \psi_j^{\uparrow}(r_j) &= r_{j+1} = \gamma(r_j), \\ \omega_j^{\uparrow}(r_j) &= s_{j+1} = r_j - \gamma(r_j), \\ \Psi_j^{\downarrow}(\psi_j^{\uparrow}(r_j), \omega_j^{\uparrow}(r_j)) &= r_j = r_{j+1} + s_{j+1}, \end{aligned} \tag{5.15}$$

where $r_1 = I$ is the transformer differential current, $s_1 = \varnothing$ and $j = 1, 2, \ldots$ By the analysis operators, the signal I is decomposed into a set of components $\{s_2, \ldots, s_j, r_j\}$; and by the synthesis operator, it can be reconstructed from $I = r_j + \sum_j s_j$. Let $\gamma(r_j) = (r_j \ominus g_j) \oplus g_j$, where $\ominus$ and $\oplus$ denote the morphological erosion and dilation, respectively, and g_j is the SE at the decomposition level j. With such a scheme, the signal I is decomposed into a set of segments that reveal the shape information of the signal.

Each half-cycle of the current signal is decomposed into several fractions, the width of which is determined by the length of the corresponding SE. The height of these fractions can therefore be viewed as the increment of the current. Prior to applying MRDS for inrush identification, current I is translated into two signals as:

$$\begin{aligned} I' &= I + c_1, \\ I'' &= -I + c_2, \end{aligned} \tag{5.16}$$

where c_1 and c_2 are pre-determined constants, so that I' and I'' are calculable by the morphological operators, e.g. $I' \geq 0$ and $I'' \geq 0$. Since inrush may reach 10–20 times of the rated current of the transformer, c_1 and c_2 can be set at values as large as 20 times the rated current.

I' is an input signal applied to deal with half-cycles that contain the peaks of I, while I'' is the other input signal to process half-cycles that contain the valleys of I. SEs g_j are simple zero-valued flat lines with lengths of l_j. Assume that f_r represents the sampling frequency of the system, then $l_j = j/f_r$ $(j = 1, 2, \ldots, \llcorner N/4 \lrcorner)$, in which N is the number of sampling points per cycle.

The process of MRDS is illustrated in Fig. 5.11. It can be seen from Fig. 5.11 that current I' is the addition of I and 20 A, and its mirror I'' is the addition of $-I$ and 50 A, which make $I' \geq 0$ and $I'' \geq 0$. MRDS begins from level $j = 1$ and ends at level $j = \llcorner N/4 \lrcorner$. A group of components, s_j, can be extracted from the currents I' and I'' using (5.15), and the height of the current increment for each of these components is measured and denoted as I_j. Assuming that the currents in Fig. 5.11 are sampled at 12 points per cycle, MRDS will iteratively run three times, and six components of s_j in total are extracted from I' and I'', respectively. I_1, I_2 and I_3 are the current increments of s_j, which are extracted from the current I'. Another group of current increments, I_1, I_2 and I_3, is obtained from I'', respectively.

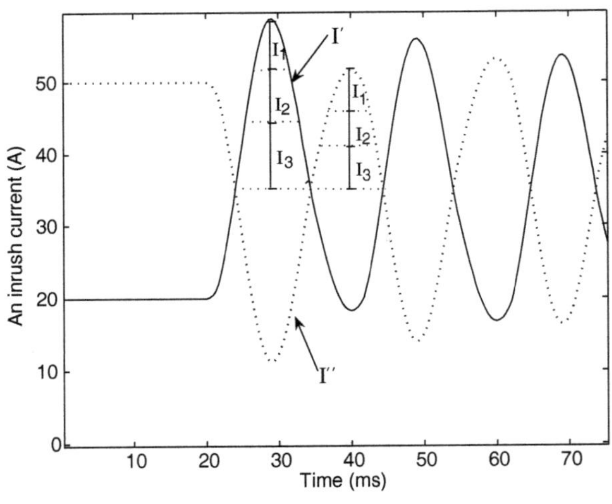

Fig. 5.11 The morphological decomposition of a current waveform

5.4.2.2 Inrush Identification

A feature criterion, σ_j, is introduced to quantify the features of the current waveform, based on the measured current increments, I_j, in comparison with the values calculated from a standard sinusoidal wave, as follows:

$$\sigma_j = \frac{I_j}{\cos((j-1)*\varphi) - \cos(j*\varphi)}, \tag{5.17}$$

where $j = 1,2,\ldots,\llcorner N/4 \lrcorner$, $\varphi = f_s * 2\pi / f_r$, and f_s is the system fundamental frequency of 50 Hz. For each half-cycle waveform of the current I, there is a group of σ_j extracted by the MRDS correspondingly. The obtained σ_j are arranged as a time series, σ_t, according to the time increment. For instance, σ_1, σ_2 and σ_3 are calculated based on the group of current increments I_j extracted from I' as shown in Fig. 5.11; σ_4, σ_5 and σ_6 are obtained from the group of I_j extracted from I''. The detailed progress is illustrated in Table 5.1.

Table 5.1 The arrangement of σ_t

I'			I''			
I_1	I_2	I_3	I_1	I_2	I_3	...
σ_1	σ_2	σ_3	σ_4	σ_5	σ_6	...

The values of σ_t reflect the slopes of the saturation curve. According to the characteristics of inrush currents discussed in the previous section, two equal increments in the core flux generate two increments in inrush of different length due to the slope variation of the saturation curve. Hence, the values of σ_t are variable from time to time if the current is an inrush. On the contrary, the values of σ_t stay the same if the current is caused by a transformer internal fault. Therefore, the variation of σ_t can be utilised for inrush identification.

5.5 Simulation Studies and Results Analysis

Various cases of internal fault and inrush currents are simulated in this section. For the former, the terminal and internal winding faults are both taken into consideration. For the latter, the inrush currents are simulated with different energising angles and different remnant fluxes in the core of the transformer. The transformer is energised from the high-voltage side and there is a circuit breaker (CB) on the low-voltage side.

5.5.1 A Transformer Model

In order to simulate internal faults in a power transformer, a three-phase transformer model is introduced in this section. The model is supplied by BCTRAN, an auxiliary routine of alternative transients program (ATP) simulation software [1]. Two matrices, **R** and **L**, are employed to describe a transformer based on positive and zero-sequence excitation, and short-circuit tests [11, 13]. For a three-phase transformer with two sets of windings as shown in Fig. 5.12, the matrices **R** and **L** are 6×6, as shown below:

$$\mathbf{R} = \begin{bmatrix} R_1 & 0 & 0 & 0 & 0 & 0 \\ 0 & R_2 & 0 & 0 & 0 & 0 \\ 0 & 0 & R_3 & 0 & 0 & 0 \\ 0 & 0 & 0 & R_4 & 0 & 0 \\ 0 & 0 & 0 & 0 & R_5 & 0 \\ 0 & 0 & 0 & 0 & 0 & R_6 \end{bmatrix}, \tag{5.18}$$

$$\mathbf{L} = \begin{bmatrix} L_1 & M_{12} & M_{13} & M_{14} & M_{15} & M_{16} \\ M_{21} & L_2 & M_{23} & M_{24} & M_{25} & M_{26} \\ M_{31} & M_{32} & L_3 & M_{34} & M_{35} & M_{36} \\ M_{41} & M_{42} & M_{43} & L_4 & M_{45} & M_{46} \\ M_{51} & M_{52} & M_{53} & M_{54} & L_5 & M_{56} \\ M_{61} & M_{62} & M_{63} & M_{64} & M_{65} & L_6 \end{bmatrix}, \tag{5.19}$$

where R is resistance, L is self-inductance, and M is mutual inductance.

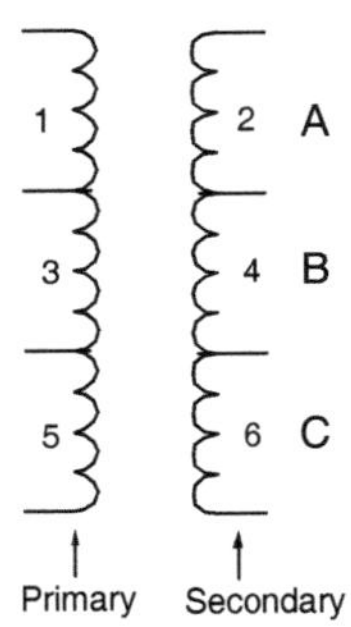

Fig. 5.12 A three-phase transformer model with two windings

In order to simulate the case of a turn-to-earth fault, the corresponding faulty winding is divided into two sub-coils as shown in Fig. 5.13. In this situation, the transformer can be modelled with two modified 7×7 matrices **R**, **L**.

In a turn-to-turn case, the faulty coil is divided into three segments as shown in Fig. 5.14. Thus, 8×8 matrices **R** and **L** are used to simulate the fault. The values of L and M in these matrices are not constant due to the variation of leakage factors between the coils. The leakage factors can be determined from the geometrical data of the transformer and the position of the fault. In the simulation of different transformer internal fault scenarios, the values of L and M are estimated and reconfigured according to the fault location and the flux leakage between the faulty coils. The detailed estimation process can be found in [11].

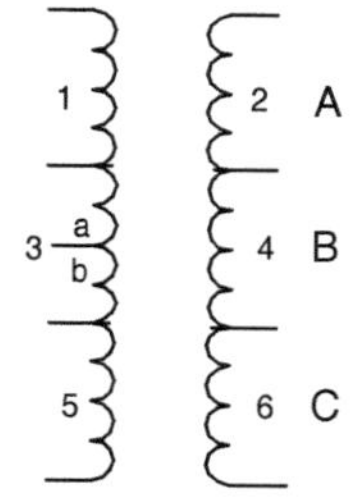

Fig. 5.13 Diagram for the study of a turn-to-earth fault

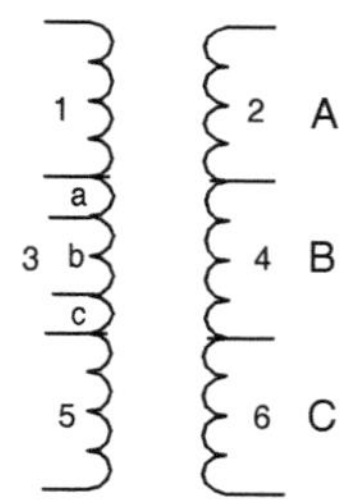

Fig. 5.14 Diagram for the study of a turn-to-turn fault

Figure 5.15 shows a transformer model that is connected to a three-phase AC power source. The characteristics and parameters of the transformer model are listed in Table 5.2. In the table, B_s and B_m represent the saturation and maximal flux, respectively. The sampling rate in this case is 48 points per cycle at a fundamental frequency of 50 Hz.

Table 5.2 The parameters of the transformer model

Transformer connection	$Y_0/\triangle$ (HV/LV)
Rated apparent power	750 MVA
Short-circuit power	4600 MVA
Rated ratio	420 kV/27 kV
Iron core	Type-96
Saturation flux density	$B_s = 1.15B_m$
Source impedance	$Z = 0.01 + 0.06j\ \Omega$

The currents in simulation studies are transformed by CTs. Non-linear effects of the saturation of CTs are considered in the studies. The connection of the CTs is illustrated in Fig. 5.15. Each CT is designed with a turns ratio of 100 : 1 based on the Jiles–Atherton model [10].

Various cases of transformer internal faults and inrush cases are simulated. For the former, both the terminal and internal winding faults are taken into consideration. For the latter, the inrush currents are simulated with different energising angles and different remnant fluxes in the core of the transformer. The transformer is energised from the high-voltage side and there is a CB located on the low-voltage

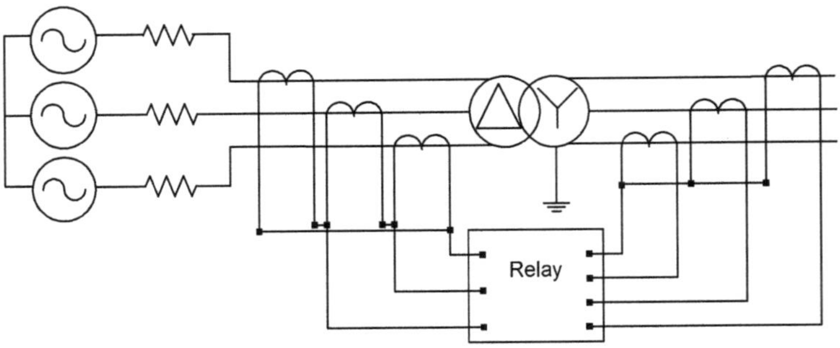

Fig. 5.15 A simulation model for transformer protection

side to control the timing of energisation. In order to demonstrate the capability of MRDS in inrush identification, two extreme cases, an inrush and an internal fault current, are selected from simulation data and dealt with by MRDS. The decomposition results of MRDS are presented for the purpose of comparison. Furthermore, a group of inrush and internal fault currents generated in experiments are analysed to validate the effectiveness of MRDS.

The simulation of inrush and internal faults is undertaken in different scenarios. Some simulation results of inrush currents are given in Table 5.3. The symbols in the table are explained as follows: B_s - the remanent flux, α - inception angles for different phases, K_{max} - the maximal ratio between the second harmonic and fundamental components, and K_{min} - the minimal ratio between the second harmonic and fundamental components.

5.5.2 Application of MDS for Inrush Detection

MDS was tested on the simulated internal fault current shown in Fig. 5.9a, as well as inrush signals, as shown in Figs. 5.16a and 5.17a. Two inrush signals, one with a second harmonic contents less than 10% and the other with the effect of CT saturation, were measured on-site from a transformer provided by ALSTOM. The internal fault and inrush signals were processed by MDS. Their partially reconstructed waveforms can be obtained by (5.11) and are shown in Figs. 5.9b, 5.16b and 5.17c, respectively. As illustrated in the partially reconstructed waveforms, only the peaks and valleys of the original signal with 0.008 s intervals exist in the output signal, in which the singularities of the inrush current are clearly enhanced; whereas as shown in Fig. 5.9b, the partially reconstructed fault current of Fig. 5.9a remains fairly symmetrical.

Figures 5.18 and 5.19 present a cycle by cycle evaluation of the waveform asymmetries, δ_I and δ_t, respectively, for the partially reconstructed signals. It is clear that in the first cycle, both δ_I and δ_t of the fault current are far below the 10% threshold, whereas the inrush currents are mostly well above 20% and may be as high as 73%.

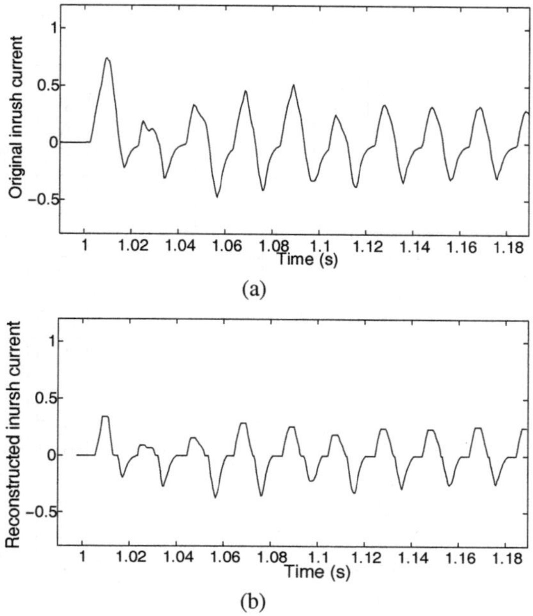

Fig. 5.16 Inrush signal 1 and its reconstruction. **a** Inrush signal 1. **b** Partial reconstruction of inrush signal 1 with $\kappa = 0.08$

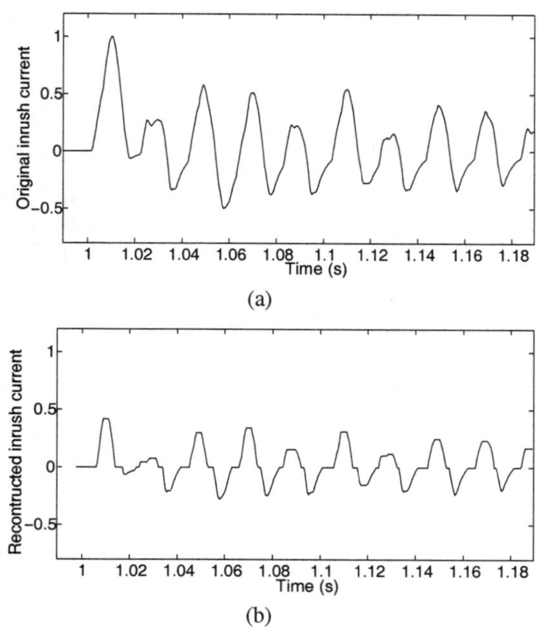

Fig. 5.17 Inrush signal 2 and its reconstruction. **a** Inrush signal 2. **b** Partial reconstruction of inrush signal 2 with $\kappa = 0.08$

Table 5.3 The second harmonic-to-fundamental ratios of inrush currents used in the simulation tests

Group	Phase	B_s	α	K_{max}	K_{min}	Group	Phase	B_s	α	K_{max}	K_{min}
	A	0.5	-120	0.33	0.32		A	0.6	30	0.14	0.13
1	B	0	-120	0.17	0.16	14	B	-0.2	-90	0.24	0.23
	C	0	120	0.40	0.38		C	-0.2	-210	0.38	0.36
	A	0.5	0	0.27	0.26		A	0.6	30	0.09	0.08
2	B	-0.3	-120	0.17	0.15	15	B	-0.6	-90	0.20	0.18
	C	-0.3	120	0.43	0.42		C	-0.6	-210	0.41	0.40
	A	0.5	0	0.22	0.21		A	0.6	0	0.19	0.18
3	B	0.5	-120	0.16	0.14	16	B	-0.6	-120	0.15	0.14
	C	-0.5	120	0.44	0.43		C	-0.6	120	0.44	0.42
	A	0.5	30	0.11	0.09		A	0.6	0	0.27	0.25
4	B	-0.5	-90	0.21	0.19	17	B	-0.2	-120	0.19	0.18
	C	-0.5	150	0.41	0.40		C	-0.4	120	0.44	0.43
	A	0.5	30	0.14	0.13		A	0.6	30	0.15	0.14
5	B	-0.3	-90	0.22	0.20	18	B	0	-90	0.32	0.31
	C	-0.2	150	0.38	0.37		C	-0.6	-210	0.40	0.39
	A	0.5	-30	0.44	0.43		A	0.6	-30	0.35	0.34
6	B	-0.5	-150	0.20	0.19	19	B	-0.6	-150	0.09	0.08
	C	-0.5	90	0.38	0.37		C	-0.6	90	0.37	0.36
	A	0.5	-30	0.46	0.45		A	0.6	-30	0.45	0.45
7	B	-0.3	-150	0.19	0.18	20	B	-0.2	-150	0.20	0.19
	C	-0.2	90	0.33	0.32		C	-0.2	90	0.33	0.32
	A	0.6	30	0.14	0.13		A	0.7	-30	0.39	0.38
8	B	-0.2	-90	0.24	0.23	21	B	0	-150	0.19	0.18
	C	-0.2	-210	0.38	0.36		C	-0.7	90	0.43	0.42
	A	0.6	30	0.09	0.08		A	0.7	0	0.16	0.14
9	B	-0.6	-90	0.20	0.18	22	B	-0.1	120	0.15	0.13
	C	-0.6	-210	0.41	0.40		C	-0.6	-120	0.43	0.41
	A	0.6	0	0.19	0.18		A	0.9	0	0.11	0.10
10	B	-0.6	-120	0.15	0.14	23	B	-0.9	120	0.14	0.13
	C	-0.6	120	0.44	0.42		C	-0.9	-120	0.43	0.42
	A	0.6	0	0.27	0.25		**A**	**0.9**	**30**	**0.08**	**0.07**
11	B	-0.2	-120	0.19	0.18	**24**	**B**	**-0.9**	**-90**	**0.14**	**0.13**
	C	-0.4	120	0.44	0.43		**C**	**-0.9**	**-210**	**0.38**	**0.37**
	A	0.6	30	0.15	0.14		A	0.9	30	0.12	0.11
12	B	0	-90	0.32	0.31	25	B	0	-90	0.34	0.33
	C	-0.6	-210	0.40	0.39		C	-0.9	-210	0.33	0.32
	A	0.6	-30	0.35	0.34		A	0.9	-30	0.32	0.31
13	B	-0.6	-150	0.09	0.08	26	B	-0.9	-150	0.19	0.17
	C	-0.6	90	0.37	0.36		C	0.9	90	0.44	0.43

One may notice that in certain cycles, e.g. the third cycle (Table 5.4), the δ_I of inrush 2 is only 4.19%. However, since the identification of inrush current is solely conducted by means of the asymmetry exhibited in the first cycle, such values should not cause any problem.

The maximal length of the SE is important in applying MDS, as it determines the length of the base for the extracted peaks and valleys, as well as their amplitudes. An optimal length of the maximal SE should allow the asymmetry in the amplitude to be

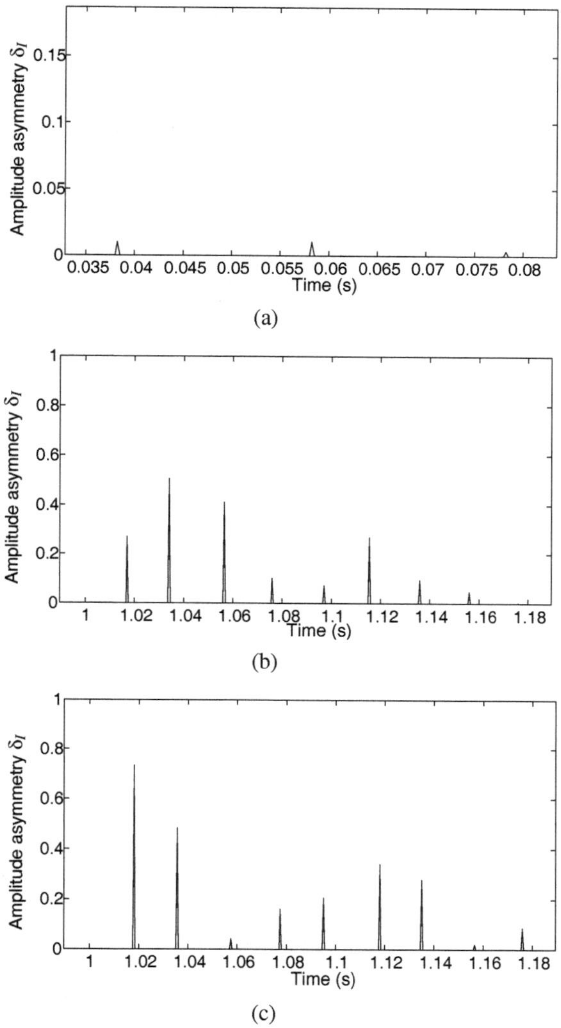

Fig. 5.18 The asymmetries on peak and valley amplitudes. **a** Internal fault. **b** Inrush 1. **c** Inrush 2

Table 5.4 δ_I of the inrush and internal fault currents

Signal	1st cycle	2nd cycle	3rd cycle
Inrush 1	27%	50.51%	41.1%
Inrush 2	73.11%	48.38%	4.19%
Internal fault	4.05%	0.2%	1.17%

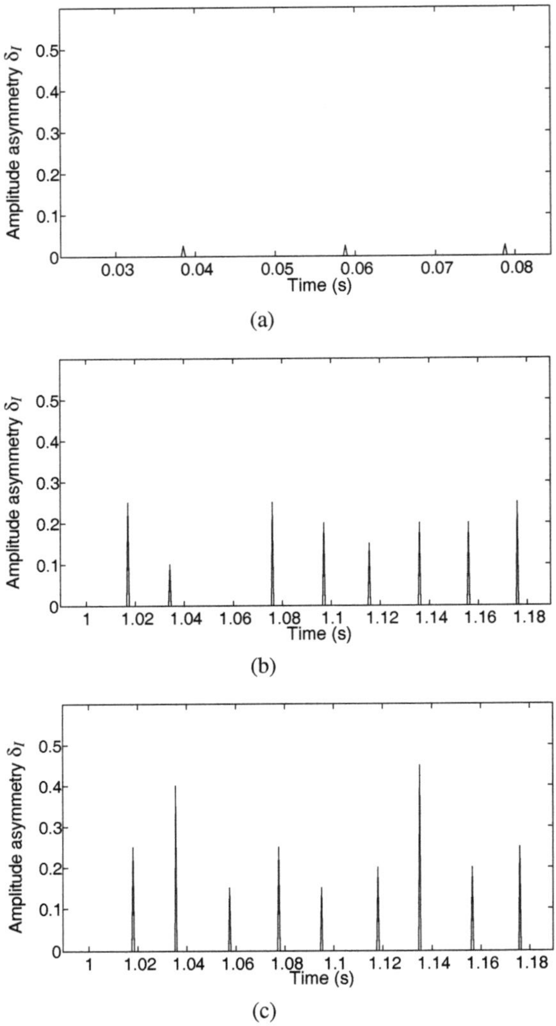

Fig. 5.19 The asymmetries of peak-to-valley intervals. **a** Internal fault. **b** Inrush 1. **c** Inrush 2

accurately extracted. A reasonable length should also be chosen to avoid excessive calculation, since the longer the SE, the more computational effort is required for the morphological operation.

The variation of δ_I with respect to the maximal SE in the first cycle is given in Fig. 5.20, for both inrush and the internal fault waveforms. It indicates that before point A at 0.008 s, the difference between inrush and fault current is not evident enough; whereas after A, the prolonged SE will lead to extra computation load. Therefore, 0.008 s is chosen as the maximal length of SE.

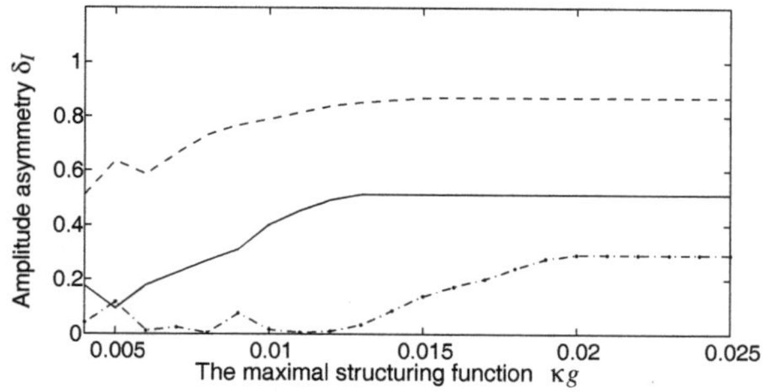

Fig. 5.20 The variation of the detected amplitude asymmetry δ_I with respect to the maximal SE

For off-line simulations where computation time is not critical, the SE may be adjusted, or may be adapted, to achieve better accuracy at the expense of computing time. For a real-time implementation, in the case where an inrush occurs at zero phase and the 0.008 s length of SE is applied, the identification requires one fundamental cycle to obtain A_{max} and A_{min}, plus 0.008 s for the dilation and erosion involved in the opening for decomposition.

5.5.3 Evaluation of MRDS for Inrush Identification

5.5.3.1 Inrush Cases

In order to validate MRDS, group 24 from Table 5.3 is selected as an extreme condition for analysing inrush currents. In this case, the remanent fluxes of the three phases in the transformer are $B_{\mathrm{ra}} = 0.9B_{\mathrm{m}}$, $B_{\mathrm{rb}} = B_{\mathrm{rc}} = -0.9B_{\mathrm{m}}$, and the inception angle of the power source at phase A is 30°. At such an inception angle, the high remanent flux shifts the flux-current trajectory far above the knee point of the saturation curve, and consequently, the magnetising current is mainly mapped from the linear part of the flux-current trajectory on the saturation curve. Therefore, the inrush currents contain comparatively low amounts of the second harmonic. The three phase inrush currents are shown in Fig. 5.21, where the second harmonic-to-fundamental ratios in the inrush currents are less than 15%, and the minimal ratio of phase A is 8%. Therefore, it may cause malfunction of relays if the threshold of the second harmonic restraint is set to be 15%, which is often the case in normal practice for transformer protection.

The feature criterion, σ_j, is calculated according to (5.17) and is represented as a time series: σ_t. The current of phase A, i_{a}, is selected to be analysed by MRDS, because it is gauged to have the lowest ratio between the second harmonic and fundamental components, which is 8%. For convenience of comparison, σ_t is normalised by σ_t/σ_1, where σ_1 is obtained by measuring the current increment of s_2, which is decomposed from the first cycle of inrush. The values of σ_t in the first cycle

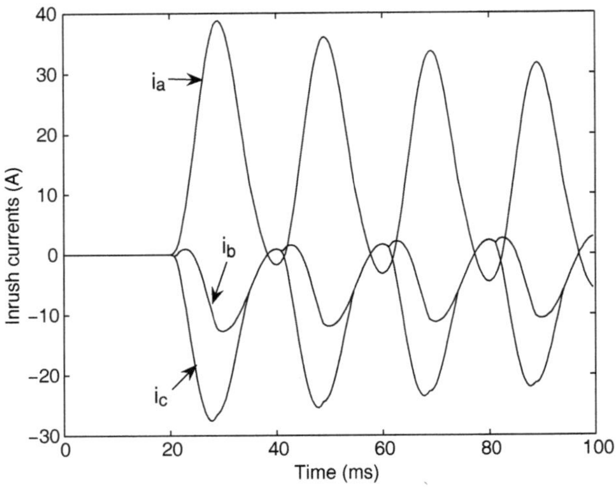

Fig. 5.21 The inrush currents of all three phases

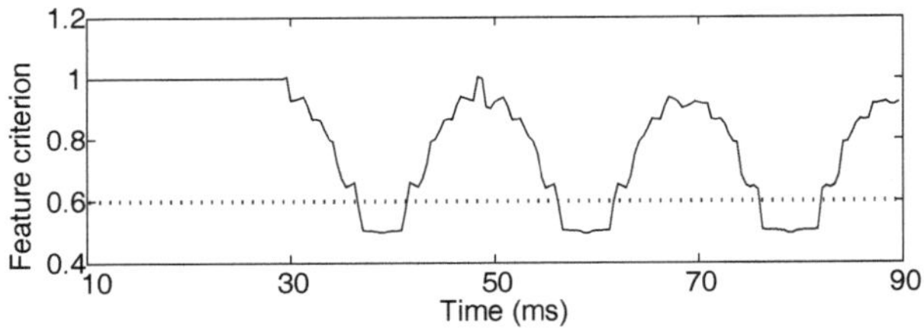

Fig. 5.22 The values of σ_t for inrush current i_a

are given in Table 5.5. Since the currents in the simulation model are sampled at 48 points per cycle, there are 24 values of σ_j obtained in each cycle according to the decomposition process of MRDS.

Table 5.5 The values of σ_t calculated for the first cycle of inrush current i_a

σ_1	σ_2	σ_3	σ_4	σ_5	σ_6
1.0000	1.0053	0.9289	0.9320	0.9373	0.9416
σ_7	σ_8	σ_9	σ_{10}	σ_{11}	σ_{12}
0.9095	0.8666	0.8683	0.8629	0.8278	0.8007
σ_{13}	σ_{14}	σ_{15}	σ_{16}	σ_{17}	σ_{18}
0.6861	0.6823	0.6949	0.7072	0.6745	0.6889
σ_{19}	σ_{20}	σ_{21}	σ_{22}	σ_{23}	σ_{24}
0.6767	0.6412	**0.5225**	0.5218	0.5216	0.5215

Figure 5.22 graphically illustrates the calculated σ_t of the inrush current i_a. The value of σ_t decreases in the first half of the cycle and then increases in the second

half of the cycle. In the second cycle, this pattern repeats, which indicates that the current is an inrush having the characteristics of a saturation curve. From Fig. 5.22, it can be seen that MRDS can identify inrush currents by measuring σ_t. If a threshold of σ_t, σ_{set}, is set at 0.6 (the optimum threshold obtained by trials), a relay block-trip signal will be generated at 38.3 ms (at the instant of σ_{21} in Table 5.5). There is a short interval after the occurrence of the inrush. Other simulation cases are considered with different residual fluxes and inception angles, and there are high second harmonic-to-fundamental ratios found in the inrush currents. The value of σ_t changes over time and shows a similar pattern.

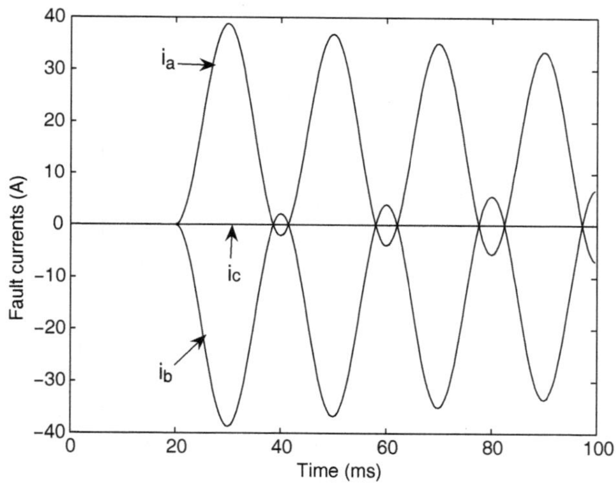

Fig. 5.23 The internal fault currents of all three phases

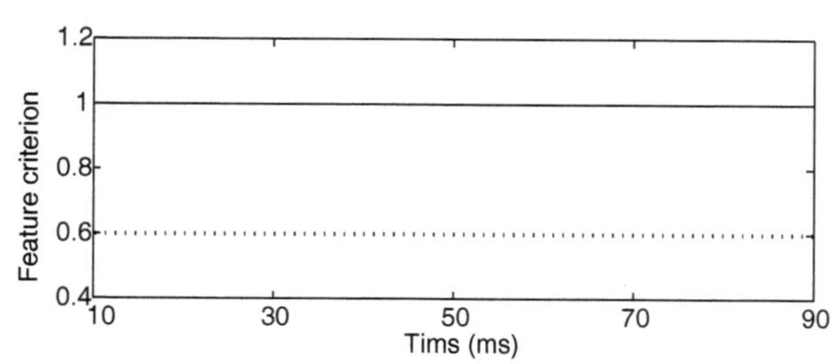

Fig. 5.24 The values of σ_t for internal fault current i_a

5.5.3.2 Internal Faults

A double-phase fault is simulated at the terminal of the high-voltage winding side of the transformer. The three phase transformer differential currents are plotted in Fig. 5.23. Since the inrush current (phase B) in Fig. 5.21 has a second harmonic-to-fundamental ratio as low as 7%, it is difficult to distinguish it from the fault current (phase B) in Fig. 5.23 based on the second harmonic components. However, the difference can be detected by MRDS. The fault current i_a in Fig. 5.23 is decomposed by MRDS, and σ_t is calculated and its curve is drawn in Fig. 5.24. Those values of σ_t corresponding to the components of the first cycle of i_a are listed in Table 5.6.

Table 5.6 The values of σ_t calculated for the first cycle of fault current i_a

σ_1	σ_2	σ_3	σ_4	σ_5	σ_6
1.0000	1.0000	1.0000	1.0000	1.0000	1.0000
σ_7	σ_8	σ_9	σ_{10}	σ_{11}	σ_{12}
0.9999	0.9999	0.9999	0.9999	0.9998	0.9998
σ_{13}	σ_{14}	σ_{15}	σ_{16}	σ_{17}	σ_{18}
1.0006	1.0006	1.0006	1.0006	1.0006	1.0006
σ_{19}	σ_{20}	σ_{21}	σ_{22}	σ_{23}	σ_{24}
1.0006	1.0006	1.0006	1.0006	1.0006	1.0005

In this case, small variations of σ_t are found during the fault period, and the values of σ_t almost remain the same. Based on the values of σ_t, the currents in Fig. 5.23 can be identified as fault currents, which are caused by a transformer internal fault. Therefore, a trip signal will be activated. More simulation studies have been carried out with various fault types and different fault locations to testify the validity of the developed scheme. The values of σ_t are no smaller than 0.65 even under the condition that the fault current has a second harmonic-to-fundamental ratio as high as 10%. Hence, an optimum threshold of σ_t can be set at 0.6.

CT saturation has a significant impact on the waveform of the secondary currents. It causes a decrement of the second harmonic components in an inrush current, which may result in a mal-operation of the relay using the second harmonic restraint approach. Figure 5.25 illustrates the distortion of an inrush current caused by CT saturation. The inrush current is simulated under the condition described in group 6 of Table 5.3. In comparison with the primary current, the secondary inrush current has weaker asymmetry in the waveform, and the zero-crossing interval of the current decreases.

MRDS is applied to decompose the secondary inrush current shown in Fig. 5.25. The normalised σ_t is calculated and plotted in Fig. 5.26. From Fig. 5.26, it can be seen that the values of σ_t vary tremendously during the period of each fundamental cycle. It decreases to a value smaller than the pre-defined threshold ($\sigma_{set} = 0.6$) at time $t = 39.2$ ms. The relay identifies the current as an inrush; thereafter, a block-trip signal is generated.

CT saturation is also considered for fault currents, the same as for inrush currents. A turn-to-earth fault is simulated with 10% of turns in phase B connected to

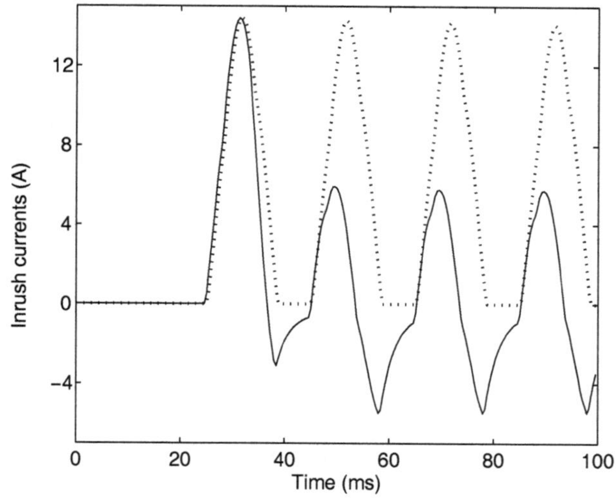

Fig. 5.25 The primary (*dashed line*) and secondary (*solid line*) inrush currents

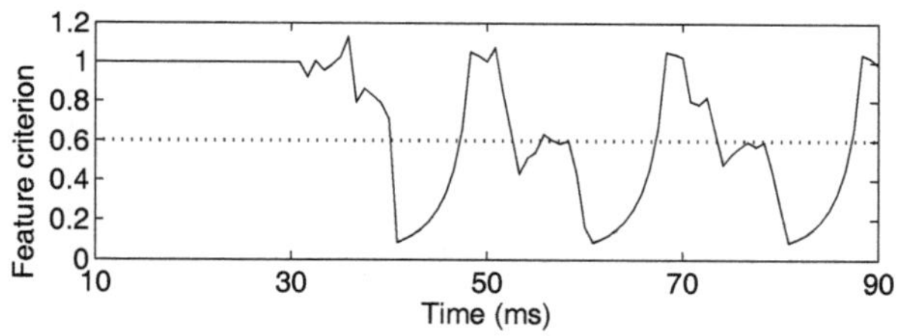

Fig. 5.26 The values of σ_t for the inrush current with CT saturation

ground at the high-voltage winding side of the transformer. The primary and secondary fault currents in phase A are shown in Fig. 5.27. The secondary fault current begins to distort from the second fundamental cycle after the fault occurrence. The current is dealt with by MRDS for inrush or fault current discrimination. The feature criterion, σ_t, is estimated and plotted against the time axis in Fig. 5.28. The variation of σ_t is not significant until the current is distorted from the second fundamental cycle, which is caused by the CT saturation. Although σ_t varies after the CT saturation, the value stays larger than the threshold of 0.6 for this case. It indicates that MRDS is still effective even when the CT is driven into light saturation. If the CT is operating in severe saturation, a method can be applied first as a pre-processing of the secondary current. Thus, the saturation part of the fault current can be detected and then compensated [105]. Afterwards, the recovered current waveform can be handled by MRDS for the discrimination of inrush and fault current.

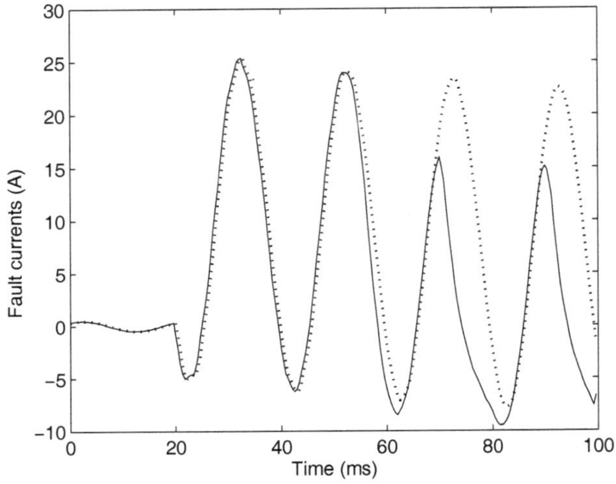

Fig. 5.27 The primary (*dashed line*) and secondary (*solid line*) fault currents

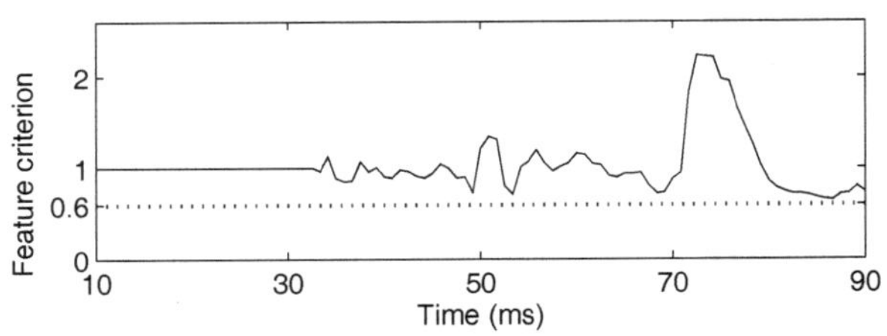

Fig. 5.28 The values of σ_t for the fault current with CT saturation

5.5.3.3 Experimental Studies

MRDS has been evaluated on an experimental fault current and inrush, respectively. The currents are measured by CTs through experiments on a transformer installed at the Power Dynamic Laboratory (Huazhong University of Science and Technology, Wuhan, China). The sampling frequency of the experimental system is 600 Hz. MRDS is only employed when a large differential current is detected by comparing the current samples at two sides of the transformer. The currents obtained from experiments are decomposed by the scheme and the calculated σ_t is plotted in Fig. 5.29.

In Fig. 5.29, the variation of σ_t of the inrush is different from that of the fault current. The minimum of σ_t is 0.08 during the first half of the cycle of the inrush from the instance of $t = 100$ ms. If the threshold, σ_{set}, is still set at 0.6 as a criterion for the transformer differential relay, the inrush current can be identified when the value of σ_t is smaller than σ_{set}. In contrast, in the case of the transformer internal fault, the value of σ_t is larger than σ_{set} during the period of the first half of the cycle. This indicates the occurrence of a fault and a trip signal will be generated after the first half of the cycle.

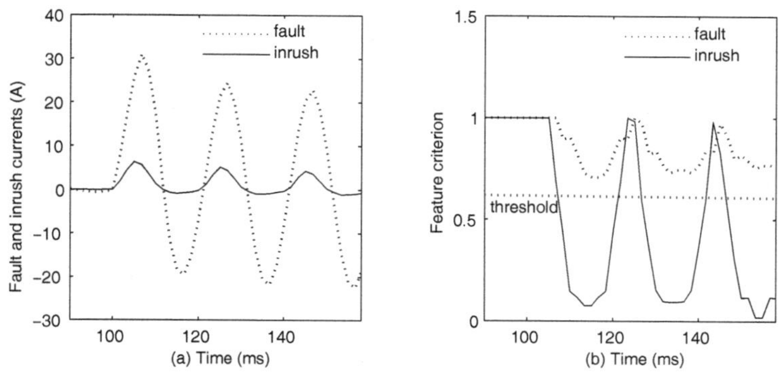

Fig. 5.29 **a** The waveforms of internal fault (*dashed line*) and inrush current (*solid line*). **b** The values of σ_t for the internal fault (*dashed line*) and inrush current (*solid line*)

From both simulation and experimental studies, it can be seen that the scheme developed succeeds in identifying inrush, while conventional methods using the second harmonic restraint may fail. The setting of the threshold, σ_{set}, at 0.6 satisfies the requirement of inrush discrimination even for the inrush with a low second harmonic component. The selection of the threshold is determined according to the magnetic characteristics of the core steel in the transformer to be protected. For modern transformers, the choice of a threshold at 0.6 guarantees both the reliability and sensitivity of the protection scheme based on our simulation and experimental tests. Therefore, the threshold can be chosen for practical transformer protection without conducting new simulations.

5.6 Further Discussion of the MM-Based Schemes for Inrush Identification

This chapter describes the MM-based schemes for the design of power transformer protection relays, to identify inrush current, taking into consideration the CT saturation condition. The schemes are fundamentally different from conventional methods, as it decomposes the input signal based on time-domain features instead of its response in the frequency domain. Since it works directly upon the geometric characteristics of the signal, there is no need for transform techniques such as Fourier, Laplace, and Hilbert transforms, nor is there a requirement for involving concepts such as frequency, convolution, effective bandwidth and ripple, which are commonly used in frequency analysis.

As shown by the results obtained, the advantages of the MM-based schemes are the following:

1. They are able to provide accurate and reliable identification of inrush in the cases when inrush current contains less than 10% of the second harmonic component. Inrush current detection is only based on time domain recognition.
2. They enable reliable operation of the protection relay in the cases that CTs are saturated.
3. There is a reduced computational complexity through the use of non-linear morphological operations with only addition/subtraction and finding local maxima/minima in calculation [122].
4. The classification schemes are a simple yet effective way of assigning an observed signal to a particular group.
5. They provide flexibility in selection of SE, which can be adjusted by modifying its shape.

Research on improving the performance of the harmonic constraint-based inrush current identification algorithms continues. However, most of these studies are either based on the transformer equivalent circuit model or require some transformer data, and this may become susceptible to parameter variations. By morphologically decomposing the signal and partially reconstructing it with a certain level of details, the asymmetrical features of the inrush waveform are exposed, whereas other irrelevant components are attenuated. With a set of simple criteria, a much better signal characterisation and a more reliable discrimination can be obtained.

5.7 Summary

Two MM-based schemes for transformer inrush current identification have been presented in this chapter. MDS discriminates the inrush current from internal fault conditions to extract the singularity associated with the asymmetric inrush waveform, while attenuating other irrelevant features. By quantifying the extracted features, the inrush current is then identified with simple criteria. This technique is simple and effective, but needs a sampling window longer than a complete fundamental cycle. Therefore, an alternative fast approach, MRDS, is introduced for distinguishing between the internal fault and inrush currents within a sampling window of 10 ms, which is half a fundamental cycle. The morphological operations decompose a current into a series of components, the features of which are extracted for inrush current detection. The schemes are evaluated using a simulated transformer model and experimental data. The simulation and experiment results presented clearly show that the MM-based schemes can accurately and promptly distinguish between internal fault and inrush currents for transformer protection. They can be used to improve the performance of differential protection relays of power transformers.

Chapter 6
Bus Protection

Abstract Differential protection relays are commonly used for bus protection. The transmission line currents injected to a busbar are measured by CTs as inputs of the relays. CT saturation can cause mal-operation of a protection relay or even prevent tripping. The waveform of the secondary current is severely distorted as the CT is forced into deep saturation when the residual flux in its core adds to the flux change caused by faults. In this chapter, a morphological lifting scheme (MLS) is introduced to extract features contained in the waveform of signal [91]. The detection of the CT saturation is accurately achieved and the points of the inflection, where the saturation begins and ends, are found by the scheme. This chapter also presents a compensation algorithm, based on the detection results, to reconstruct healthy secondary currents. MLS and the compensation algorithm are evaluated on a sample power system. The simulation results clearly indicate that they can successfully detect and compensate the distorted secondary current of a saturated CT containing residual flux.

6.1 Introduction

Bus protection is one of the most important parts of power plant and sub-station automation. It uses the combination of CTs and relays to detect faults occurring within the busbar zone of a busbar sub-station, and initiates tripping of all those circuit breaks and isolates these faults. Therefore, it is called 'bus-zone protection'. The failure of bus protection would be one of the worst cases in power systems. In this case, a large number of electric components connected to the bus would be invalidated, which would cause a large area of power loss and affect the stability of the whole power grid.

Considerable thought has been given by protection engineers regarding the need for bus protection. If dedicated bus protection is not used, clearing of bus faults is performed by the relay that protects the backup protective zones of the transmission lines terminating at the bus. This may be satisfactory in terms of the reliability of fault clearing, but it is generally rather slow and the fault would cause more damage to the mechanical structure of the bus and the surrounding equipment than would be the case with fault clearing. Therefore, high speed protection is necessary to reduce the damage that might be caused by a bus fault. The clearing time of a bus fault should always be faster than that of backup protection of the lines connected to the bus.

A number of circuit breakers can be disconnected by tripping a single bus, which may possibly lead to line outages. In this case, on the one hand, an adequately designed bus layout may reduce the potential of line outages, and on the other hand, the reliability of bus relays is also important. There are several prominent causes of security failures in bus relays, which include faults in the relay circuits, a lack of proper selectivity, mechanical shock to the relay, incorrect relay settings, and maintenance personnel errors. The security of bus protection can be improved by appropriate station design for the bus arrangement. Therefore, the bus protection can be separated into multiple protection zones for smaller bus sections. When a bus fault occurs, the resulting bus outage will affect a smaller portion of the entire station and result in a smaller system disturbance. If a bus protection security failure occurs, the effect on the power system will be less severe if the protected bus connects fewer system components.

6.2 Bus Differential Protection

Differential protection is one of the most common methods for bus protection. The concept of differential protection is based on Kirchhoff's law. Most of the bus protection systems used are current differential systems and are based on the 'current law', which requires all the currents entering the bus to be summed to zero. However, in the case of an internal fault, the sum of currents measured at the CT locations will not be zero, and then a tripping action should be taken.

Bus differential protection can be performed using over-current relays. Since the over-current relay may sense a rather high current for an external fault, it will be necessary to set the relay pickup to a high value in order to avoid false tripping.

Bus differential relaying can also employ a high impedance bus differential scheme, which is basically a current differential protection scheme using over-voltage relays. This type of bus protection is particularly well suited for situations where a large number of circuits are connected to the bus.

6.3 Current Transformers for Bus Protection

AC-type protection relays are actuated by current and voltage supplied by CTs and VTs. These transformers provide insulation against the high voltage of the power circuit, and also supply the relays with quantities proportional to those of the power circuit. The proper application of CTs and VTs involves the consideration of several requirements such as mechanical construction, type of insulation, and ratio in terms of primary and secondary currents or voltages.

CTs used for protection relays can be made from different materials. The iron core toroidal CT is usually chosen for relaying in higher-voltage circuits. This type of CT consists only of an annular-shaped core with a secondary winding. It is built into equipment such as circuit breakers, power transformers, generators or switchgear. The core is arranged to encircle an insulating bushing through which a power conductor passes.

CT accuracy is always of concern. For relaying purposes, it is necessary to determine the phase-angle error of a CT. The load on the secondary side of a CT is generally of such a highly lagging power factor that the secondary current is practically in phase with the exciting current. Hence, the effect of the exciting current on the phase-angle accuracy is negligible. Furthermore, most relaying applications can tolerate a phase-angle error for metering purposes. If the ratio error is tolerable, the phase-angle error can be neglected.

A CT is burdened when an external load is applied to the secondary current of the CT. The burden is preferably expressed in terms of the impedance of the load and its resistance and reactance components. The term 'burden' is applied not only to the total external load connected to the terminals of a CT but also the elements of the load. The publications of manufacturers give the burdens of individual relays, from which the total CT burden can be calculated together with the resistance of interconnecting leads. The CT burden impedance decreases as the secondary current increases, because of saturation in the magnetic circuits of relays and other devices.

6.4 Current Transformer Problems

CTs are widely used for both measurement and protection purposes in power systems. Distorted secondary currents, due to CT saturation, can cause measurement error and malfunction of protection relays [82]. Thus, the CT saturation problem must be considered for designing a protection relay.

Some methods for detecting CT saturation onset have been suggested in [23, 78, 109]. These methods only successfully detect the CT saturation based on some given assumptions such as that the current collapses to zero as soon as the CT saturates and the residual flux at the beginning of the calculation is zero. These assumptions make the methods' applicability limited since they are not always satisfied in real systems.

An impedance-based CT saturation detection scheme for busbar differential protection was introduced in [42]. It uses the first-order differential equation of a secondary current, at the relay position, to detect the beginning and end points of inflection during the period of CT saturation. As the value of the first-order derivative at the instant next to the first point of inflection beginning is not large enough, another saturation detection algorithm was developed, using the third-order derivative of the secondary current [77]. However, compared with methods that only require knowledge of the first-order derivative, it needs a longer calculation time to obtain more distinct results, and suffers more from signal disturbances.

The approach of reducing the impact of CT saturation is to reconstruct the secondary current waveform, using a compensation algorithm. For the purpose of reconstruction, a function with given CT parameters was applied to approximately represent the non-linear core characteristics of a specific model of CT [78]. Another method, using ANN [150], attempts to learn the non-linear characteristics of CT magnetisation and restructure the waveform based on the learned characteristics. However, due to the variations of CT saturation characteristics and secondary burdens, these methods cannot be universally applied in various situations.

In order to tackle the problem of CT saturation, advanced signal processing techniques are sought to deal with the shapes of the signals. WT has been recognised as a powerful tool for analysing the transient shape distortion, but it cannot be adaptively and universally applied to different situations since it is linear in its original form. The linearity limitation makes it unable to deal with the CT saturation detection since it is difficult to distinguish the beginning and end points of inflection during the period of CT saturation from the disturbances of noise. The emergence of the lifting scheme [137] has made non-linear extensions of WT possible, since it provides a useful way to construct non-linear wavelets in the decomposition process [69]. In [30], an adaptive lifting step using a non-linear selection criterion was introduced by Claypoole et al. They also combined linear and non-linear lifting steps (based on a median operator), and discussed its applications in image compression and noise removal [31, 32, 33]. As a non-linear branch of image and signal processing methods, MM may be used to construct a family of non-linear wavelets, called morphological wavelets. Its major difference from the classical linear wavelet is that the linear signal analysis filter in the lifting scheme is replaced by the morphological operators.

MLS, based on morphological wavelets, is introduced in this chapter as a powerful analysis tool to detect CT saturation. Afterwards, a compensation algorithm is developed to reconstruct the distorted secondary current. The developed scheme is a signal decomposition method based on the processing of the shape of secondary current waveform. The method has the ability to extract important features from the distorted signal when CT saturation occurs, and it does not require any preset parameters. Based on the detection of CT saturation characteristics, a real-time compensation of the distorted secondary current can be achieved.

This chapter firstly addresses the issues of the CT characteristics, saturation detection and compensation. Afterwards, MLS for CT saturation detection and a compensation algorithm, based upon the detection results, are explained in detail. Their performance, tested under various faults and CT conditions, is evaluated based on a simple power system model.

6.5 Saturation of Current Transformers

The iron core toroidal CT is widely used in the electric power industry for the measurement of current. This type of CT is not a linear transducer due to the characteristics of the iron core. Different levels of saturation occur in almost all CTs depending on the magnitude of the fault current being measured.

Figure 6.1 shows a simplified equivalent circuit of a CT for transient analysis, where Z_m is the excitation impedance, and R_2 and L_2 are the secondary inductance and resistance, respectively. Normally, the value of Z_m is very large when the iron core runs at a low magnetic density. Therefore, the secondary current I_2 can be perfectly transformed from the primary current I_1.

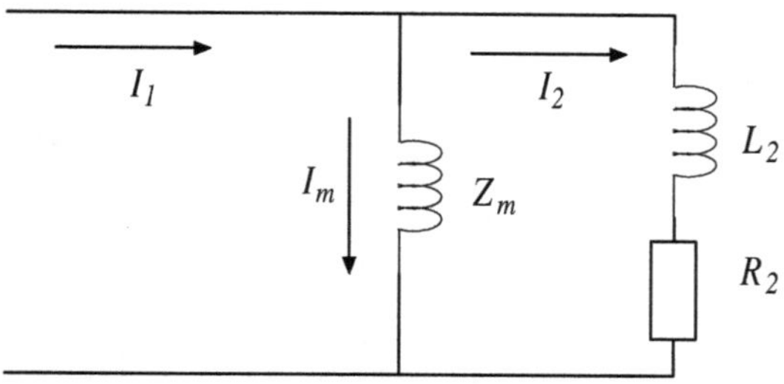

Fig. 6.1 The simplified equivalent circuit of a CT for transient analysis

When a system fault occurs, the fault current through the primary circuit of a CT can be considered as an exponential component superimposed on a sinusoidal component, given by:

$$i_1(t) = Ae^{-\alpha t} + B\sin\phi_t, \tag{6.1}$$

where A, B, α and ϕ_t are the initial values of the exponential component and the amplitude of the sinusoidal component, a decaying coefficient, and a fault inception angle, respectively.

With a large exponential component added to the primary current, the magnetic density is promptly boosted to saturation, which will cause Z_m to descend. Thus the magnetising current I_m through the iron core will increase and distort the secondary current I_2. When the CT runs under heavy saturation, Z_m is almost zero and all the current flows through the excitation coil, which makes the secondary current I_2 decrease to almost zero. This will lead to gross measurement errors or malfunctions in relay tripping.

An important factor that affects CT saturation is the core. Increasing the size of core or using a core material that supports large flux densities can avoid CT saturation. However, both options can affect the cost of transformer applications. A practical method used to correct CT saturation needs to be implemented in the existing device without increasing the cost. To this end, an MLS for CT saturation detection and a compensation algorithm are investigated to deal with the CT saturation problem, which seriously influences the operation of various protection relays in power systems. Figure 6.2 is a block diagram showing how the scheme and algorithm can be implemented for real-time detection and compensation of CT saturation. The secondary current I_2, transformed from the primary current I_1 by a CT, is sampled into a discrete-time sequence of values, $i_2(t)$, by a data-acquisition module. If a sample of $i_2(t)$ is determined within a saturation portion of the fault current waveform by the CT saturation detection scheme, a compensated sample will be generated by a compensation algorithm. Finally, the compensated currents are supplied to protection relays.

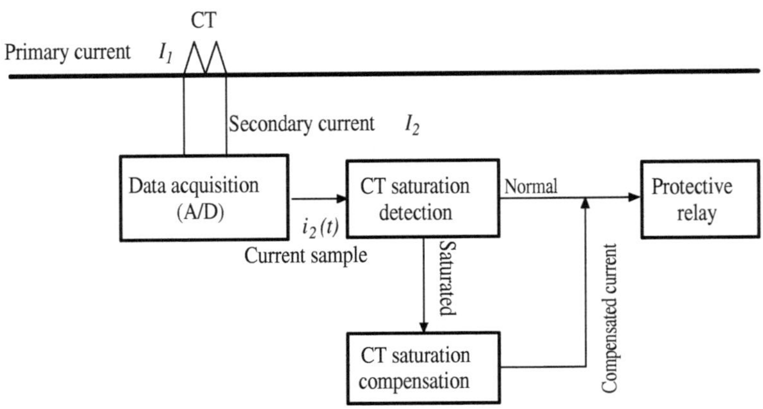

Fig. 6.2 Real-time detection and compensation of CT saturation

6.6 A Morphological Lifting Scheme for Detection of CT Saturation

New morphological operators are introduced to make a morphological scheme applicable in the detection of CT saturation in power systems. Since the basic power system signals are composed of sinusoids, a symmetric SE, k_m, with the characteristics of a sinusoid can be employed as:

$$k_m = [\cos(2m+1)\varphi, \ldots, \cos\varphi, \cos\varphi, \ldots, \cos(2m+1)\varphi], \tag{6.2}$$

where $\varphi = 2\pi f t$, f is the power system frequency and t is the sampling time instant. The total length of k_m is $2m$, where m is a positive integer.

According to the designed SE, a new type of dilation and erosion are defined as:

$$(x \oplus k_m)(n) = \max_{n-v\in\mathscr{D}_x, v\in\mathscr{D}_k} \{x(n-v)/k_m(v)\}, \tag{6.3}$$

$$(x \ominus k_m)(n) = \min_{n+v \in \mathscr{D}_x, v \in \mathscr{D}_k} \{x(n+v)/k_m(v)\}, \tag{6.4}$$

where x is the signal to be processed and $\mathscr{D}_x$ and $\mathscr{D}_k$ denote the domains of x and k_m, respectively.

As mentioned in Sect. 2.4.2, in the lifting scheme, the original data stream of signal $x(n)$ is split into two data streams: $x_e[n]$ and $x_o[n]$. Therefore, the detail signal d is obtained as the prediction residue lifted by:

$$d[n] = x_o[n] - P(x_e[n]), \tag{6.5}$$

where P represents the morphological prediction operator, which is the mean of dilation and erosion results using different SEs with different lengths. It is defined as:

$$P(x_e) = \frac{1}{2l} \sum_{i=1}^{l} (x_e \oplus k_i + x_e \ominus k_i), \tag{6.6}$$

where k_i is an SE with a length of $2i$ $(i = 1, 2, \ldots, l)$ as defined in (6.2). When $i = l$, there are total $2l$ even samples (x_e) used in prediction, with l samples on the left and l on the right. If only one pair of even samples on the left and on the right are used, the prediction is denoted as:

$$P(x_e) = \frac{1}{2}(x_e \oplus k_1 + x_e \ominus k_1), \tag{6.7}$$

where $k_1 = [\cos\varphi, \cos\varphi]$. This MLS is highly adaptive for the sinusoidal signal compared with the classic morphological operators or the classic WTs. If the original signal x is a pure sinusoid in the fundamental frequency, the detail signal d should always have a zero output, since the steady component of the original signal is eliminated and the high-frequency components are left only in prediction residuals.

Since the current abruptly changes when CT saturation sets in, the MLS can successfully detect the saturation period based on the distorted secondary current. In a non-saturation period, the detail signal should have an output value near zero; however, it stays at a large value whilst the CT is in saturation.

6.7 A Compensation Algorithm for Distorted Secondary Current

The distorted secondary current can be compensated based on the detection of the CT saturation period. As shown in Fig. 6.1, the relationship between the secondary current $i_2(t)$ and the core flux $\lambda(t)$ of CT can be expressed as follows:

$$\frac{d\lambda(t)}{dt} = Ri_2(t). \tag{6.8}$$

Let t_0 represent the starting time of saturation. Therefore, the integration of λ from t_0 to t is:

$$\lambda(t) - \lambda(t_0) = R\int_{t_0}^{t} i_2(t)\mathrm{d}t. \tag{6.9}$$

The core flux $\lambda(t)$ is determined by the magnetisation current $i_m(t)$ according to the magnetising curve. For the purpose of simplification, the saturated section of the magnetising curve is regarded as a straight line with a slope of K. The unit of K is H/S^{-1}. Hence, the magnetising current can be obtained as:

$$K(i_m(t) - i_m(t_0)) = R\int_{t_0}^{t} i_2(t)\mathrm{d}t. \tag{6.10}$$

Since the value of $i_m(t)$ at time t_0 is near zero, $i_m(t)$ is calculated as:

$$i_m(t_0 + n\cdot\triangle t) = \frac{R}{K}\sum_{j=0}^{n} i_2(t_0 + j\cdot\triangle t), \tag{6.11}$$

where $\triangle t$ represents the time interval between two samples and n denotes the nth sample instant counted from time t_0. In (6.11), the value of R/K can be deduced by a linear interpolation for the first point of the magnetising current $i_m(t_0 + \triangle t)$:

$$\begin{aligned} i_m(t_0 + \triangle t) = {} & [i_2(t_0) + (i_2(t_0) - i_2(t_0 - \triangle t))] \\ & - i_2(t_0 + \triangle t); \end{aligned} \tag{6.12}$$

$$\frac{R}{K} = i_m(t_0 + \triangle t)/(i_2(t_0) + i_2(t_0 + \triangle t)). \tag{6.13}$$

Then the secondary current referred to as the distorted current, $i_2'(t)$, is reconstructed by summing the measured secondary current $i_2(t)$ with the estimated magnetising current $i_m(t)$ sample by sample. The compensation process is illustrated in Fig. 6.3.

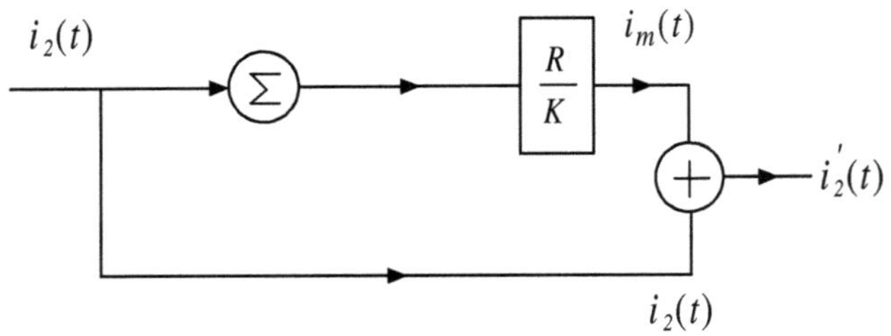

Fig. 6.3 The compensation process

6.8 Case Studies

A simple power system is considered to demonstrate the effectiveness of MLS for CT saturation detection and the compensation algorithm. It consists of two sources and a single line as shown in Fig. 6.4. S1 and S2 are equivalent sources, with Z_{S1} and Z_{S2} as their equivalent impedances, respectively. The phase-to-phase voltage of the sources is 120 kV and the system frequency is 50 Hz. The total length of the transmission line is 200 km. The CT saturation is simulated when a fault occurs near the CT location with a sampling frequency of 5 kHz.

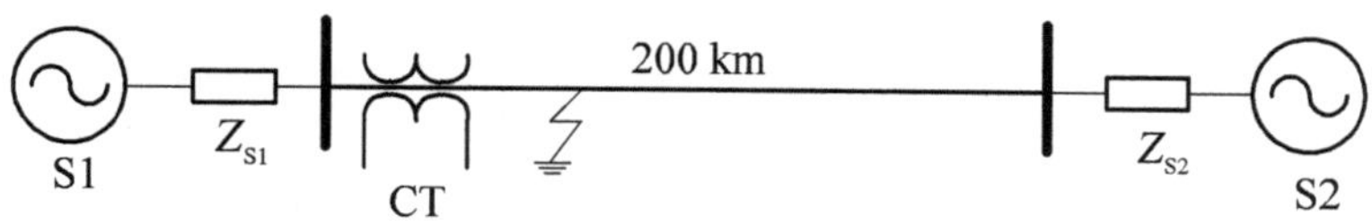

Fig. 6.4 A simple power system model

A model of a saturable CT was employed for case studies. The 2000 : 5 tap was selected from a C400 class of CT. The magnetising branch is represented internally as a non-linear inductor. Since the residual flux in the core needs to be considered, the Type-96 inductor was employed for its ease of initialising the residual flux in the CT core. To use the model, the CT hysteresis characteristics should be evaluated in advance. The saturation knee point is set at 10 p.u. (per unit) flux. Different faults were simulated by changing the residual flux from −80% to 80% of the flux at the saturation point.

In the following discussion, the secondary current is firstly processed using MLS and then reconstructed by the compensation algorithm. An SE, k_2, with a length of 4, e.g. a moving window of seven samples was used in the computation process. The scheme and algorithm were tested for a variety of cases. The cases presented cover a wide range of fault conditions, including the changes of residual flux, fault inception angle and fault types. The simulation results clearly indicate that all distorted sections of the secondary currents can be satisfactorily detected and then reconstructed using the method in a wide variety of cases.

6.8.1 Tests Undertaken at Different Levels of Residual Flux

Figure 6.5 shows the detecting result in the case of a remanence of −80% residual flux in the CT core. The polarity of the remanence is opposite that of the DC offset, which results in a slight distortion beginning at the second cycle of the fault secondary current, as shown in Fig. 6.5. In Figs. 6.5–6.8, dashed and solid lines represent the primary and distorted secondary current, respectively.

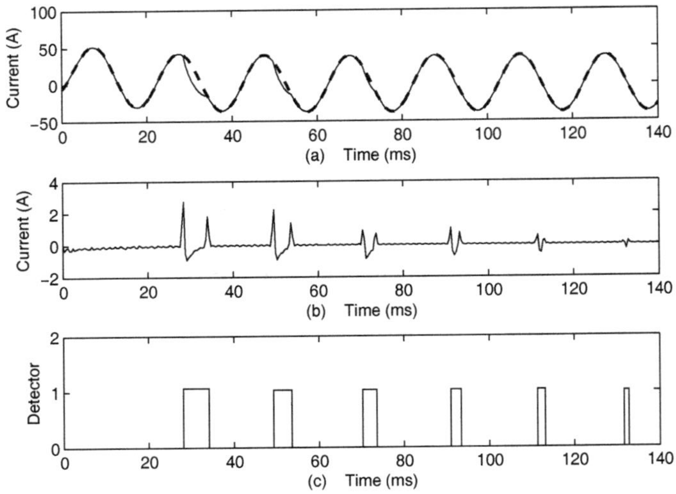

Fig. 6.5 The case of a −80% residual flux with a single phase-to-ground fault. **a** The primary (*dashed line*) and distorted secondary current (*solid line*). **b** The detail signal d extracted from the secondary current. **c** Saturation detection output

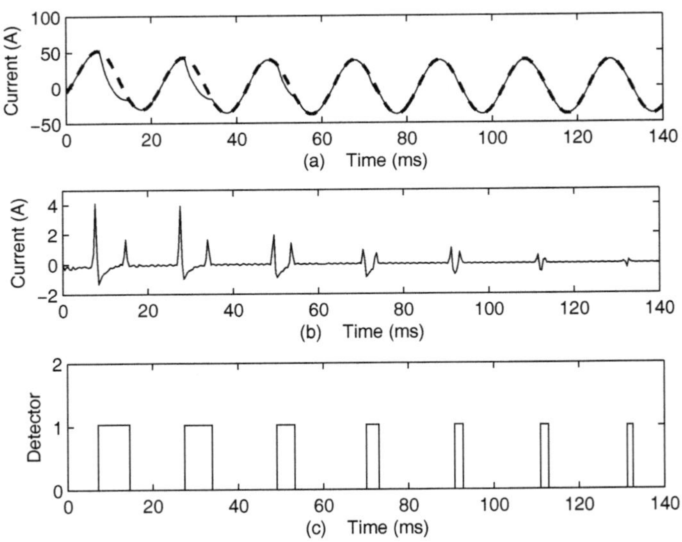

Fig. 6.6 The case of a 0% residual flux with a single phase-to-ground fault. **a** The primary (*dashed line*) and distorted secondary current (*solid line*). **b** The detail signal d extracted from the secondary current. **c** Saturation detection output

The detail signal d is extracted by the scheme as shown in Fig. 6.5b. It is composed of six periods of the wave with a similar pattern. The absolute values in these periods are much larger than zero, which indicates that the CT runs in a saturated state. Figure 6.5c is the output of the detector in a binary mode, where '1' means that the CT is saturated, while '0' represents an unsaturated state. The detection results clearly show that MLS successfully detects the interval of the saturation.

The extent of the distortion gradually becomes more severe as the remanence increases from a negative value to a positive value. The case shown in Fig. 6.6 has 0% remanence within the CT core. The distortion of the secondary current starts at the first cycle after the fault occurs. Figure 6.6c clearly shows that MLS accurately detects the distorted sections of the original secondary current waveform.

Figure 6.7 illustrates the distortion associated with 80% residual flux, which is the maximum flux possible from a practical CT. In this situation, the transit flux has a very short effective excursion before saturation occurs. It is apparent from the figure that the distortion of the secondary current is the most severe in comparison with other situations. It begins only a few milliseconds after a fault occurs.

The results indicate that MLS detects the interval of saturation even when the secondary current is severely distorted due to a high residual flux.

The orientation of the distortion will change if the fault inception angle is adjusted to make the fault current contain a negative DC offset. The secondary current will have upward distortion as shown in Fig. 6.8a, where the CT has 80% residual flux with an opposite fault inception angle. Compared with the previous cases, the detail signal obtained has similar shapes with a different polarity. The CT saturation interval is also successfully detected by MLS, as shown in Fig. 6.8c.

Tests on various faults and CT conditions are simulated and the results obtained indicate that the CT saturation detection algorithm accurately estimates the secondary current irrespective of the magnitude of the primary time constant, the fault types, the DC component, or the residual flux.

6.8.2 Compensation Results

The distorted secondary currents in the above cases can be reconstructed, based upon the detection results, using the compensation algorithm. To evaluate the performance of the algorithm, the transient error for the compensated secondary currents is calculated at each sampling point as follows:

$$\text{transient error }(\%) = \frac{K_\mathrm{t} \cdot i_2'(t) - i_1(t)}{I_\mathrm{r}} \times 100(\%), \tag{6.14}$$

where I_r is the rated magnitude of the primary short-circuit current and K_t is the turns ratio of the CT. In this chapter, $K_\mathrm{t} = 400$, since the 2000 : 5 tap was selected for the C400 class CT used in the simulation study.

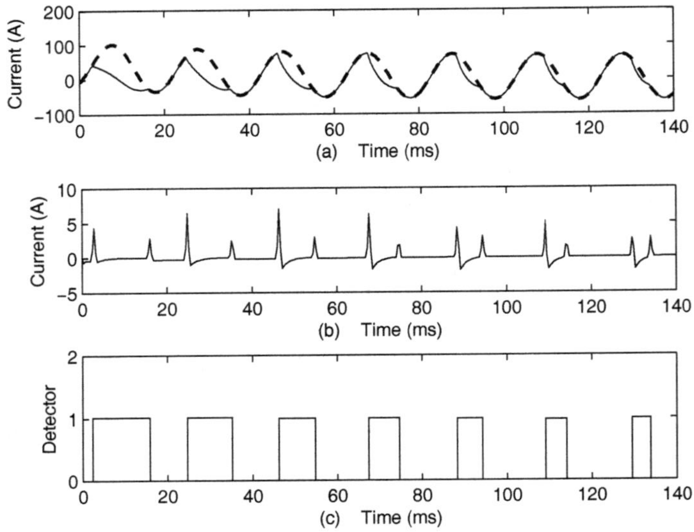

Fig. 6.7 The case of an 80% residual flux with a three phase-to-ground fault. **a** The primary (*dashed line*) and distorted secondary current (*solid line*). **b** The detail signal *d* extracted from the secondary current. **c** Saturation detection output

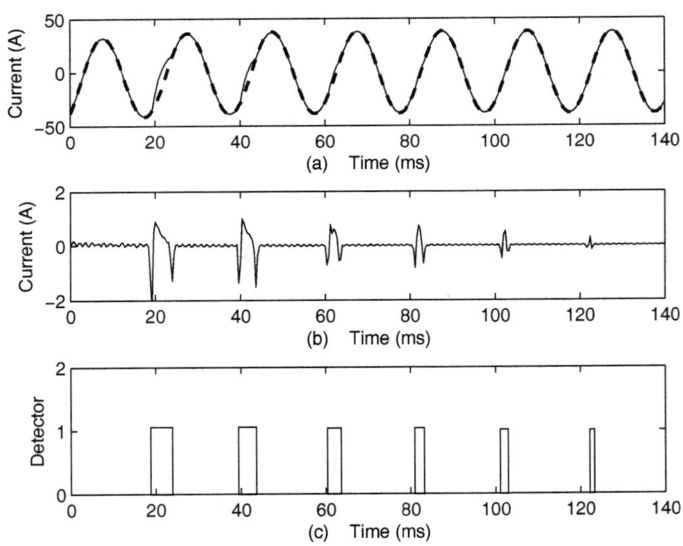

Fig. 6.8 The case of an 80% residual flux with a single phase-to-ground fault. **a** The primary (*dashed line*) and distorted secondary current (*solid line*). **b** The detail signal *d* extracted from the secondary current. **c** Saturation detection output

Figure 6.9a is the compensation result of the distorted secondary current in Fig. 6.6a, which is reconstructed based on the detection results indicated in Fig. 6.6c. The maximum transient error shown in Fig. 6.9b is less than 3%, which indicates that the compensation algorithm can accurately reconstruct the distorted secondary current during the period of CT saturation.

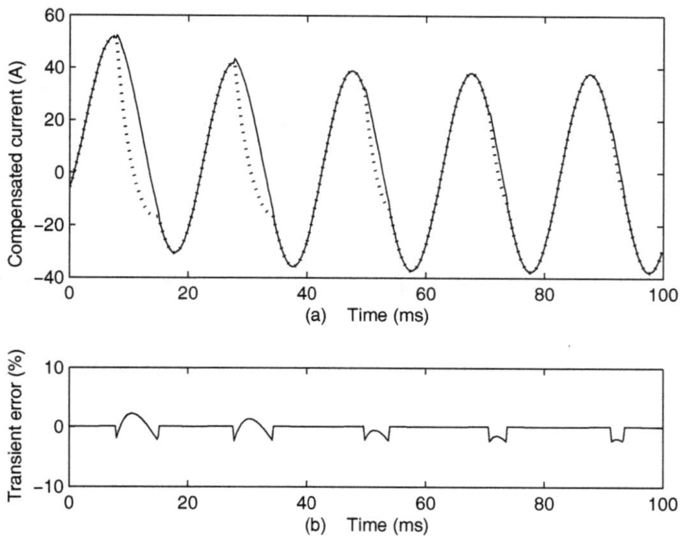

Fig. 6.9 The compensation results for the secondary current in Fig. 6.6. **a** The saturated (*dashed line*) and compensated (*solid line*) secondary current. **b** Transient error

Figure 6.10 illustrates the compensation result for the case of an 80% residual flux with a single phase-to-ground fault, which was indicated in Fig. 6.8. The solid line in Fig. 6.10a is the compensated secondary current obtained sample by sample at a sampling interval of 0.2 ms. The transient error is calculated and shown in Fig. 6.10b. The whole compensation process does not require any additional parameters and can be fully implemented in real time.

MLS and the compensation algorithm for CT saturation are much superior to other existing methods, as most existing methods are based upon the assumption that there is no residual flux in the core of CT. Obviously, this assumption is not practical and it is not used in our approach. In other words, without knowledge of the residual flux, other existing methods can by no means be compared with the method developed herein.

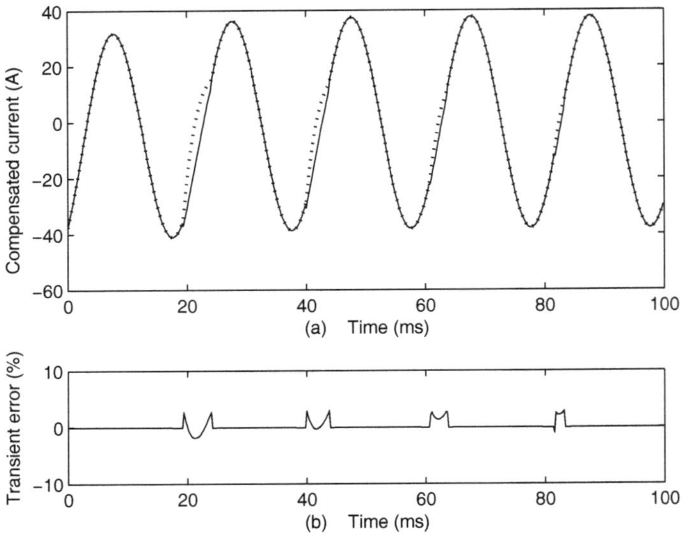

Fig. 6.10 The compensation results for the secondary current in Fig. 6.8. **a** The saturated (*dashed line*) and compensated (*solid line*) secondary current. **b** Transient error

6.9 Summary

This chapter introduced a promising method using an MLS with a compensation algorithm for the saturation detection and the compensation of the secondary current of CTs. The method is based on the non-sinusoidal characteristics of the secondary current waveform during the saturation period. The detail level signal, after lifting by the scheme, has a zero or non-zero output, when the CT runs at normal or saturated conditions, respectively. It can accurately distinguish the distorted section from the healthy section in the current waveform. The distorted secondary current can be compensated in real time once saturation begins by using the compensation algorithm.

The effectiveness of MLS and compensation algorithms has been tested on different cases on a simple power system. The faults used in the evaluation of the method are simulated in a wide range of conditions, including the residual flux in the CT core changing from -80% to 80% and the fault inception angle ranging from 0° to 180°. Despite the wide variations, our method is able to detect the period of secondary current wave distortion accurately in all the test cases and then compensate the distorted secondary current sample by sample in real time. Since no pre-set parameters are needed for MLS and compensation algorithms, they can be easily used to incorporate the existing relays to improve their action speed without using conventional blocking schemes.

Chapter 7
Ultra-High-Speed Protection

Abstract Chapters 7 and 8 are concerned with the study of transient-based ultra-high-speed (UHS) protective relaying. In this chapter, we concentrate on the directional protection of the transmission lines and the positional protection is explicated in the next chapter. The directional protection relay derives the direction discriminants of a fault occurring on a transmission line and determines the phase of the fault. MMGA, presented in Sect. 2.4.4, is employed to extract the wavefront of a transient. As we will see in this chapter, the fault-generated transient features contained in the voltage and current signals are extracted and recognised by the maxima and polarities of their MMG outputs. Based on MMG of the signals, two composite relaying signals are fabricated to indicate the direction of the fault occurring on the transmission line.

7.1 Introduction

UHS protection has been recognised for a long time as an effective way of improving the transient stability of power systems [24]. Many studies have been undertaken, especially during the 1970s and 1980s. Great contributions ascribed to the development and establishment of travelling wave-based UHS protection techniques include various basic principles, field tests and implementations. Compared with the conventional protection techniques developed from measurements of fundamental components, travelling wave-based protection has the advantage of fast response and immunity to the influence of power swing, CT saturation and the series compensation capacitor. However, it has the drawback that when a fault occurs at a low voltage inception angle, the fault-generated travelling wave is too faint to detect [19, 72]. In order to overcome this drawback, transient-based protection has been proposed. It fully utilises the fault-generated high-frequency components, including travelling waves and the components caused by fault arcing, which spans a wide bandwidth of frequencies of up to several hundreds kHz even under the condition of a low fault inception angle. Sharing the advantages of travelling wave-based protection, transient-based protection has been drawing growing attention. The development of transient-based protection relies upon the study of the relationship between transient current, voltage, time and distance, which is fully described by the travelling wave equations [14]. Therefore, it is essential to accurately extract the characteristics of the transient components comprising superimposed quantities and noises generated by a fault and its concomitant arcing respectively.

Most transient-based protection relays are designed for double-ended lines, that is, a directional comparison arrangement is accomplished by linking two terminals of a transmission line with a communication channel. Non-communication protection relays have also been developed by observing only the measurements of a transmission line at one terminal [19]. For transient-based protection relays, the most important matter is how a sequence of wavefronts in measured fault-generated transients can be extracted. It is difficult for conventional algorithms, such as FT or digital filters, to achieve both high-frequency component calculation in the frequency domain and an accurate location in the time domain, while processing a measured fault-generated transient signal, since the transient signal is non-stationary, and many non-periodic components are superimposed on the fundamental frequency waveform.

WT uses a short window for the analysis of the high frequency content of a signal and a long window for low frequency content. Thus, WT is able to identify the details of localised transients of the signal in contrast to FT and to give a time-frequency representation of the signal, as it can provide the time and frequency information simultaneously. Its application has been widely investigated for fault location [95], high-impedance fault detection [64], as well as power quality detection and classification [46]. The advent of WT has given a great impetus to investigate the possibility of improving UHS protection [27, 37, 152]. Since WT has notable capabilities of detection and localisation of transients, it is possibly capable of extracting the wavefronts of the transients. However, its capabilities are often significantly degraded due to the existence of noise riding on the transients, as the spectrum of the noise coincides with that of the transient components. Most of the noise is due to transposed transmission lines, adjacent busbars, and instrumental transducers [27]; and it is unable to exclude the effect of the noise by means of various filters without affecting the performance of WT. Moreover, the transient is prone to be contaminated by noise due to high sampling frequencies. The wavefronts are too faint to be distinguished from the noise or interferences, especially in the case of large ground-fault resistances or low inception fault angles, where the gradients of the wavefronts are quite small. The accuracy of the WT detection result will be affected in such a noisy environment. Another major disadvantage for WT to be applied to UHS protection is that it involves a large amount of computation during such a small sampling interval, which would significantly complicate the design and degrade the reliability of the relay system.

This chapter introduces MMGA for UHS directional protection and demonstrates its capability of extracting the characteristics of the transient voltage or current signals [145]. In contrast to FT and WT, this MM-based approach concentrates on the shapes of the signal and signal components completely in the time domain. Furthermore, its fast and simple calculation within a very short shifting data window will be observed. All these merits are desirable in the application of UHS protective relaying.

7.2 Principles of UHS Protection

A transient-based protection relay utilises fault-generated high-frequency wave components to trigger fast relaying actions. When a fault occurs on a transmission line, both the voltage and current signals are transmitted on the transmission line at nearly the speed of light. The voltage or current of a transient is captured by a specially designed transient detector and then converted into an aerial mode signal by modal transform. The fault line section on transmission lines is identified by comparing polarities of the transmitted sequence of wavefronts of the transient, and the fault location is determined by calculating the time tags among the sequence of wavefronts.

7.2.1 Transient-Based Directional Protection

The basic principle of directional protection is addressed with reference to a 400 kV transmission line system as shown in Fig. 7.1. A directional protection relay is situated at the end of section P near busbar S. The reference forward direction is defined at the relaying position, as current flowing from busbar S into the protected section P.

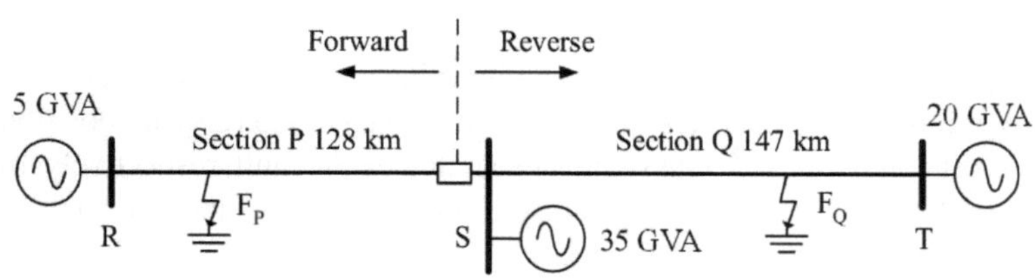

Fig. 7.1 Single-line diagram of a 400 kV power transmission line system

Considering an arbitrary fault occurring somewhere on a transmission line, according to the principle of superposition, it is known that such a disturbance is equivalent to the sudden injection of a fictitious voltage source at the location of the fault, equal in amplitude and opposite in sign to the voltage at the fault location in a pre-fault steady state. Appendix A gives a detailed analysis of the basic principles of the fault-generated transient. Assuming that a single-phase lossless transmission line has distributed constant inductances and capacitances, the relationship between transient signals, generated by a fault and measured at any point propagating along the transmission line, is given by the D'Alembert formulae [75] as follows:

$$\Delta v(x,t) = F_1\left(t - \frac{x}{c}\right) + F_2\left(t + \frac{x}{c}\right), \tag{7.1}$$

$$\Delta i(x,t) = \left(F_1\left(t - \frac{x}{c}\right) - F_2\left(t + \frac{x}{c}\right)\right)/Z_0, \tag{7.2}$$

which implies that the transient travels with a constant amplitude at a speed of c, with the combination of a forward and a backward direction of x from the fault location, which are represented by functions $F_1(t-x/c)$ and $F_2(t+x/c)$, respectively. Z_0 is known as the surge impedance. However, at points of discontinuity, part of the incident wave is reflected back along the line and the rest is transmitted into and beyond the discontinuity. Using the superscripts of F, B and T to indicate the incident (forward), reflected (backward) and transmitted wave at the point of discontinuity respectively, according to the derivation in Appendix A.3, they can be expressed as:

$$\Delta v^{\mathrm{F}}(t) = \Delta v_{\mathrm{F}}(t - x/c), \tag{7.3}$$

$$\Delta v^{\mathrm{B}}(t) = K_{\mathrm{R}}\Delta v^{\mathrm{F}}(t), \tag{7.4}$$

$$\Delta v^{\mathrm{T}}(t) = K_{\mathrm{T}}\Delta v^{\mathrm{F}}(t), \tag{7.5}$$

where $\Delta v_{\mathrm{F}}(t)$ is a fictitious voltage at the fault location, K_{R} and K_{T} are reflection and refraction coefficients, respectively, and their values are determined by the equivalent surge impedances from the point of discontinuity on both sides.

The relaying signals adopted in this scheme, initially introduced in [70] and [141], are given as:

$$S_{1\mathrm{S}} = \Delta v_{\mathrm{S}}(t) - R_1\Delta i_{\mathrm{S}}(t), \tag{7.6}$$

$$S_{2\mathrm{S}} = \Delta v_{\mathrm{S}}(t) + R_1\Delta i_{\mathrm{S}}(t), \tag{7.7}$$

where R_1 is a surge replica resistance and its value is arranged to closely match the line surge impedance Z_0; $\Delta v_{\mathrm{S}}(t)$ and $\Delta i_{\mathrm{S}}(t)$ observed at the relaying position S are the transient voltage and current signals generated by the fault, respectively. In order to illustrate how to use the directional discrimination to detect the fault, a theoretical derivation is given in detail as follows.

7.2.2 Faults in the Forward Direction

Figure 7.2 shows the equivalent superimposed circuit with a fault in forward direction, at a distance x_{FP} from relaying position S and the propagation of fault-generated transient waves along the transmission lines. The superimposed voltage $\Delta v_{\mathrm{FP}}(t)$ and associated current $-\Delta v_{\mathrm{FP}}(t)/Z_{0\mathrm{P}}$ at fault location $\mathrm{F_P}$ caused by the injection of an equivalent fictitious source, are to travel at a speed c towards the line terminals. Here, Z_{0P} is the line surge impedance of section P. The transient relaying signals observed at relaying position S can be derived by applying (7.1)–(7.4) as follows:

$$\begin{aligned}\Delta v_S(t) &= \Delta v_S^F(t) + \Delta v_S^B(t)\\ &= \Delta v_{FP}(t - x_{FP}/c) + K_R^F \Delta v_{FP}(t - x_{FP}/c)\\ &= (1 + K_R^F)\Delta v_{FP}(t - x_{FP}/c),\\ \Delta i_S(t) &= (-\Delta v_S^F(t) + \Delta v_S^B(t))/Z_{0P}\\ &= (-1 + K_R^F)\Delta v_{FP}(t - x_{FP}/c)/Z_{0P}.\end{aligned}$$

Hence,

$$S_{1S} = 2\Delta v_{FP}(t - x_{FP}/c), \tag{7.8}$$
$$S_{2S} = 2K_R^F \Delta v_{FP}(t - x_{FP}/c). \tag{7.9}$$

In this case, the reflection coefficient is determined by:

$$K_R^F = (Z_{0ST} - Z_{0P})/(Z_{0ST} + Z_{0P}),$$

where Z_{0ST} is the equivalent surge impedance viewed from the point of discontinuity S outwards.

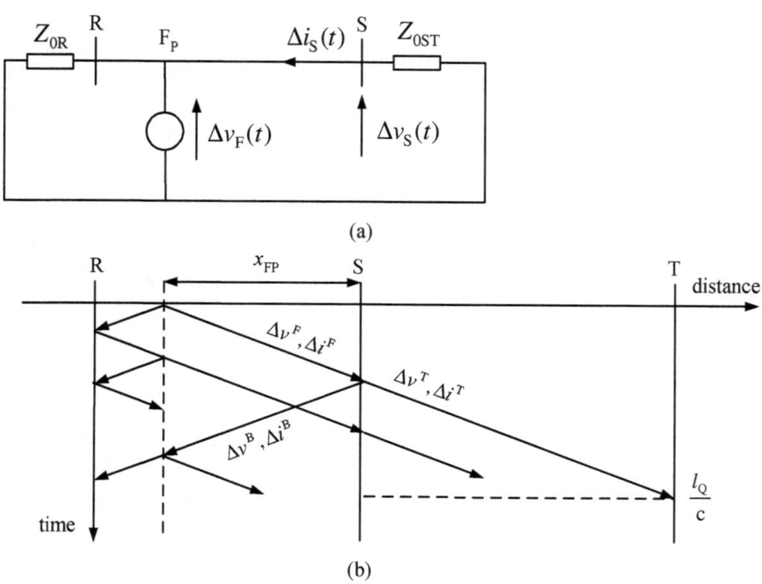

Fig. 7.2 a The equivalent superimposed circuit under a fault in the forward direction. **b** Its Bewley–lattice diagram

7.2.3 Faults in the Reverse Direction

For a fault occurring at point F_Q on section Q in the reverse direction as indicated in Fig. 7.3, only the transmitted transients can be observed at relaying position S, and their travelling direction is forward. The relaying signals can be derived similarly using (7.1)–(7.3) and (7.5) as follows:

$$\Delta v_S(t) = \Delta v_S^T(t) = K_T^B \Delta v_{FQ}(t - x_{FQ}/c),$$
$$\Delta i_S(t) = \Delta v_S^T(t)/Z_{0Q} = K_T^B \Delta v_{FQ}(t - x_{FQ}/c)/Z_{0Q},$$

where x_{FQ} is the distance between busbar S and point F_Q, as shown in Fig. 7.3. Hence,

$$S_{1S} = 0, \tag{7.10}$$
$$S_{2S} = 2K_T^B \Delta v_{FQ}(t - x_{FQ}/c), \tag{7.11}$$

where the refraction coefficient is:

$$K_T^B = 2Z_{0SR}/(Z_{0SR} + Z_{0Q}),$$

where similarly Z_{0SR} is the equivalent terminating surge impedance, and Z_{0Q} is the line surge impedance of section Q.

From the foregoing analysis, the following summary can be given:

- The direction discriminability to a fault with reference to the relaying position is definite with comparison between magnitudes of transient relaying signals S_{1S} and S_{2S}.
- Equations (7.8) and (7.9) representing the relaying signals for a forward fault, and (7.10) and (7.11) for a reverse fault will hold true for a confirmation time T_{set}, by setting $T_{set} = \min\{2l_P/c, 2l_Q/c\}$, where l_P and l_Q are the line length of section P and Q, as indicated in Figs. 7.2 and 7.3, respectively. After time T_{set}, the direction discriminability will be lost due to the arrival of the transient reflected from the far end of the adjacent line section.
- In practice, taking the stray capacitance of the busbar into account, the terminating equivalent surge impedance is not equal to the line surge impedance [16], i.e. $K_R^F \neq 0$ and $K_T^B \neq 0$.
- With extension to a three-phase transmission system, the phase variations of currents and voltages can be decomposed into modal components by Clarke modal transformation [29]. An aerial mode will be utilised in this scheme.
- In practice, a threshold, δ, should be preset to determine the zero value of the relaying signal functions in (7.6) and (7.7). Therefore, fault discrimination can be made according to the following criteria:

$$\begin{cases} |S_{1S}| > \delta, & |S_{2S}| > \delta, & \text{for a forward fault;} \\ |S_{1S}| < \delta, & |S_{2S}| > \delta, & \text{for a backward fault;} \\ |S_{1S}| < \delta, & |S_{2S}| < \delta, & \text{for no fault.} \end{cases}$$

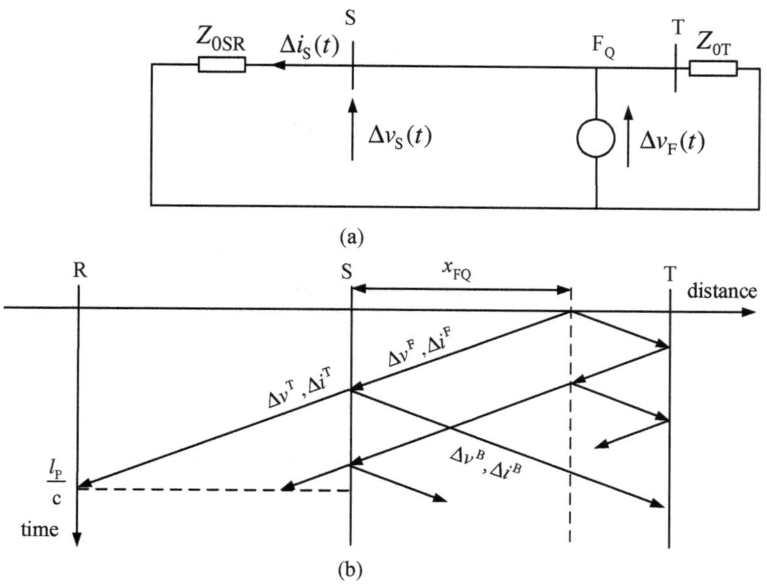

Fig. 7.3 **a** The equivalent superimposed circuit under a fault in the reverse direction. **b** Its Bewley–lattice diagram

7.2.4 Fault Phase Selection

Single-pole auto-reclosure has long been recognised as an effective means of improving system security and reliability. Its benefits are particularly apparent in applications where economic and stability considerations preclude the use of three-phase auto-reclosure. Single-pole auto-reclosure applications involve tripping only the fault phase under single-phase-to-ground fault conditions and initiating three-phase auto-reclosure for other types of fault. Reliable fault phase selection is thus vitally important in order to avoid either tripping of the incorrect phase or unnecessary three-phase tripping, thereby undermining system security and stability. Moreover, a concomitant requirement of phase selectors is high-speed operation, since the selection process must be completed in the immediate post-fault period before breaker opening [20].

After a forward fault is detected, the fault type can be selected. Depending on the fault type, a single-phase trip or a three-phase trip can be issued with the reception of a clearing signal from the opposite line terminal. Applying the Clarke modal transformation together with (7.6) and (7.7), the following discriminants can be derived for fault phase selection:

$$\begin{aligned}
S_{\text{ph1}} &= (\Delta v_{\text{B}}(t) - \Delta v_{\text{C}}(t)) - R_1(\Delta i_{\text{B}} - \Delta i_{\text{C}}(t)),\\
S_{\text{ph2}} &= (\Delta v_{\text{C}}(t) - \Delta v_{\text{A}}(t)) - R_1(\Delta i_{\text{C}} - \Delta i_{\text{A}}(t)),\\
S_{\text{ph3}} &= (\Delta v_{\text{A}}(t) - \Delta v_{\text{B}}(t)) - R_1(\Delta i_{\text{A}} - \Delta i_{\text{B}}(t)),\\
S_{\text{ph4}} &= (2\Delta v_{\text{C}}(t) - \Delta v_{\text{A}}(t) - \Delta v_{\text{B}}(t)) - R_1(2\Delta i_{\text{C}} - \Delta i_{\text{A}}(t) - \Delta i_{\text{B}}(t)),\\
S_{\text{ph5}} &= (2\Delta v_{\text{A}}(t) - \Delta v_{\text{B}}(t) - \Delta v_{\text{C}}(t)) - R_1(2\Delta i_{\text{A}} - \Delta i_{\text{B}}(t) - \Delta i_{\text{C}}(t)),\\
S_{\text{ph6}} &= (2\Delta v_{\text{B}}(t) - \Delta v_{\text{C}}(t) - \Delta v_{\text{A}}(t)) - R_1(2\Delta i_{\text{B}} - \Delta i_{\text{C}}(t) - \Delta i_{\text{A}}(t)),
\end{aligned}$$

where $\Delta v_{\text{A}}(t)$, $\Delta v_{\text{B}}(t)$, $\Delta v_{\text{C}}(t)$ and $\Delta i_{\text{A}}(t)$, $\Delta i_{\text{B}}(t)$, $\Delta i_{\text{C}}(t)$ are transient three-phase voltage and current signals observed at relaying position S. Table 7.1 lists the criteria for fault phase detection. If any discriminant among $|S_{\text{ph1}}|$ to $|S_{\text{ph6}}|$ exceeds the threshold during a period of confirming time, a single-phase or a three-phase trip will be generated accordingly.

Table 7.1 The fault phase selection scheme

Fault phase	$\lvert S_{\text{ph1}}\rvert$	$\lvert S_{\text{ph2}}\rvert$	$\lvert S_{\text{ph3}}\rvert$	$\lvert S_{\text{ph4}}\rvert$	$\lvert S_{\text{ph5}}\rvert$	$\lvert S_{\text{ph6}}\rvert$
A	$<\delta$	$>\delta$	$>\delta$	$>\delta$	$>\delta$	$>\delta$
B	$>\delta$	$<\delta$	$>\delta$	$>\delta$	$>\delta$	$>\delta$
C	$>\delta$	$>\delta$	$<\delta$	$>\delta$	$>\delta$	$>\delta$
AB	$>\delta$	$>\delta$	$>\delta$	$<\delta$	$>\delta$	$>\delta$
BC	$>\delta$	$>\delta$	$>\delta$	$>\delta$	$<\delta$	$>\delta$
CA	$>\delta$	$>\delta$	$>\delta$	$>\delta$	$>\delta$	$<\delta$
ABC	$>\delta$	$>\delta$	$>\delta$	$>\delta$	$>\delta$	$>\delta$

7.3 UHS Directional Protective Relaying

Transient directional protection is frequently used to recognise the difference between currents being supplied in forward and backward directions on a transmission line. Firstly, the wavefronts of the transients at both ends of the protected transmission line are processed using the algorithm of directional protection. Then, each pair of detected wavefronts of each phase of currents at two line terminals are put together to compare their polarities. Consequently, a fault can be either an internal or an external fault, and the fault type can also be determined. The simulation model of the transmission system is shown in Fig. 7.1, which is composed of two sections of 400 kV transmission lines connected in series. The forward current direction is defined from busbars to the transmission line. The directional protection relay is installed at busbar S to protect section P.

For all the simulation studies in this section, the current and voltage signals observed at relaying position S are sampled at the rate of 1 MHz. They are transformed into aerial mode and the simulation studies are carried out on the aerial mode current and voltage signals by applying a quadratic MMGA (level $a = 2$) with a flat SE of length 8. The results of processing the aerial mode signals by MMGA are denoted by $\rho_{\rm i}^2$ and $\rho_{\rm v}^2$, respectively, which represent the transient components. Substituting $\Delta i_{\rm S}$ and $\Delta v_{\rm S}$ in (7.6) and (7.7), the fault direction can be determined by comparing the magnitudes of $S_{1\rm S}$ and $S_{2\rm S}$.

7.3.1 A Solid Fault in the Forward Direction

Under the condition of a solid phase-A-ground fault with a nearly 90° fault inception angle, at a distance of 48 km from busbar S on section P, the three-phase fault currents and voltages observed at measurement point S are shown in Figs. 7.4a and b, respectively. The obtained aerial mode current and voltage signals, $i_{\rm S2}$ and $v_{\rm S2}$, as well as their associated quadratic MMGs, $\rho_{\rm i}^2$ and $\rho_{\rm v}^2$, are shown in Figs. 7.5a and b, respectively. It can be noted from the figures that the distinct transient characteristics of fault current and voltage signal are well presented in their locations and polarities by the quadratic MMGs, whereas the steady state components are depressed sufficiently. Examining the relaying signals $S_{1\rm S}$ and $S_{2\rm S}$ in Fig. 7.5c, it can be concluded that firstly, there is a fault occurring somewhere on the transmission line system with confirmation of spikes representing the propagation of fault-generated transients; secondly, a forward direction to a fault can definitely be determined with respect to their magnitudes. A tripping signal is then issued as $|S_{1\rm S}| > \delta$ and $|S_{2\rm S}| > \delta$.

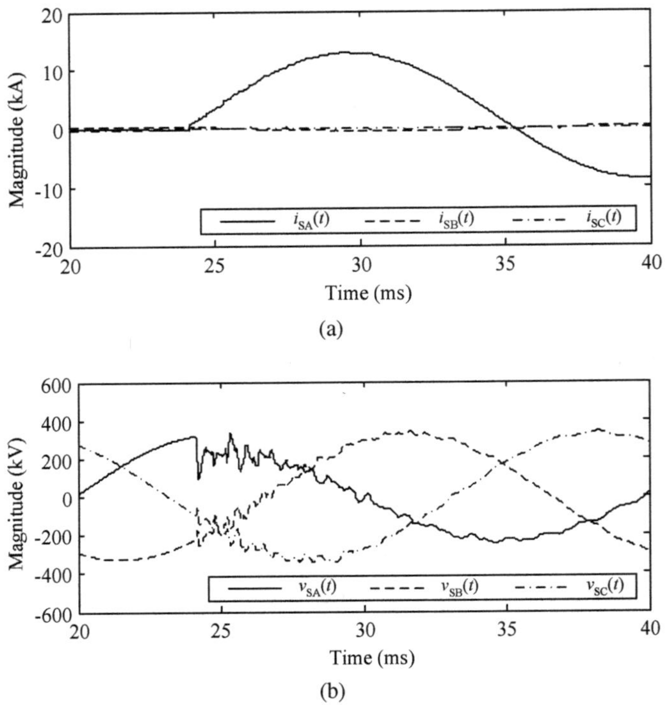

Fig. 7.4 The three-phase currents and voltages measured at busbar S under a solid phase-A-ground fault with a nearly 90° fault inception angle at 48 km from busbar S in section P. **a** Three-phase currents measured at relaying position S. **b** Three-phase voltages measured at relaying position S

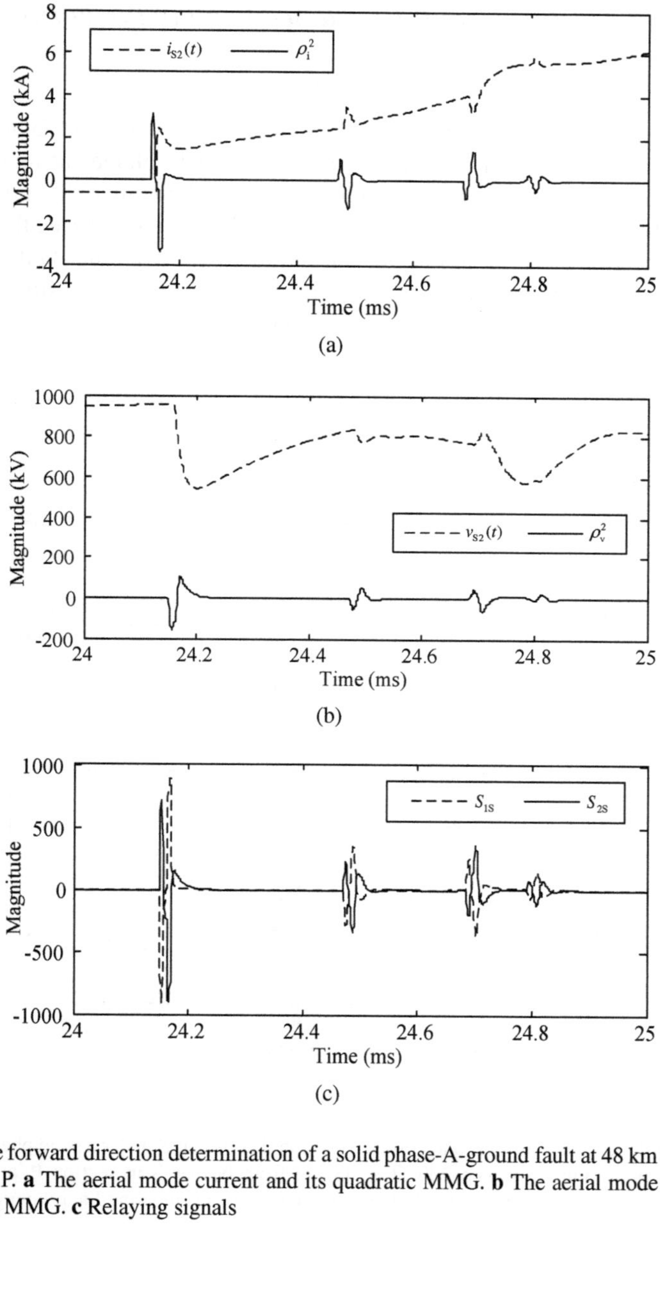

Fig. 7.5 The forward direction determination of a solid phase-A-ground fault at 48 km from busbar S in section P. **a** The aerial mode current and its quadratic MMG. **b** The aerial mode voltage and its quadratic MMG. **c** Relaying signals

7.3.2 A High Ground-Fault Resistance Fault in the Forward Direction

Under the same fault condition as described previously, the ground-fault resistance is raised to 200 Ω. The three-phase fault currents and voltages at relaying position S are shown in Fig. 7.6. The quadratic MMGs of the fault aerial mode current, voltage and relaying signals are shown in Fig. 7.7. The results indicate that the magnitudes of quadratic MMG diminish with the increase of ground-fault resistance. However, the transient characteristics presented by quadratic MMG still remain sufficient enough to enable the determination of fault direction correctly.

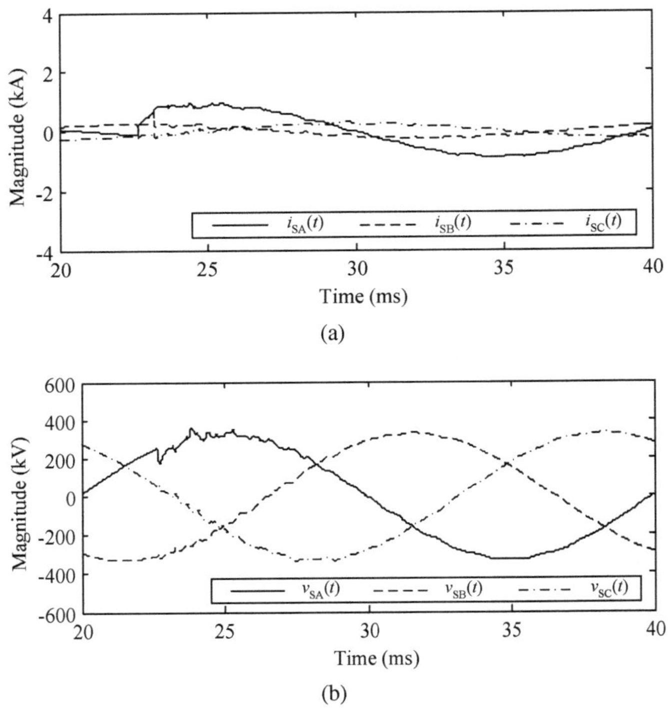

Fig. 7.6 The three-phase currents and voltages measured at busbar S under a solid phase-A-ground fault with ground-fault resistance 200 Ω at 48 km from busbar S in section P. **a** Three-phase currents measured at relaying position S. **b** Three-phase voltages measured at relaying position S

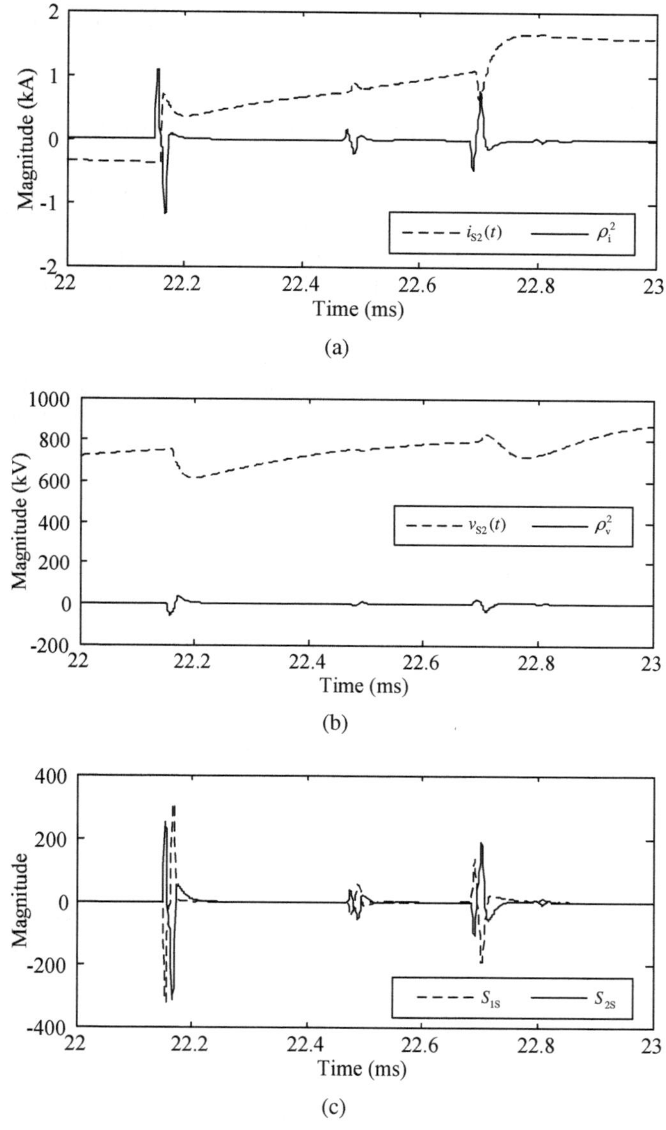

Fig. 7.7 The forward direction determination of a phase-A-ground fault with ground-fault resistance 200 Ω at 48 km from busbar S in section P. **a** The aerial mode current and its quadratic MMG. **b** The aerial mode voltage and its quadratic MMG. **c** Relaying signals

7.3.3 A Fault with Low Inception Angle in the Reverse Direction

This case aims to demonstrate the high sensitivity of MMGA. A fault with an inception angle of nearly 0° at a distance of 47 km from busbar S in section Q is simulated. The fault occurs at the time of 0.02 s. The current and voltage waveforms of the fault are shown in Figs. 7.8a and b, respectively. As can be seen, the fault-generated transients are faint. When it travels away from the fault location, the magnitudes and rates of the transient decline as the point on the voltage wave approaches a zero crossing. Nevertheless, the high-frequency transient signals generated by fault arcing are still attainable and detectable [71, 73]. Figures 7.9a and b illustrate the quadratic MMGs of the transient aerial mode obtained from the current and voltage signals, respectively. From the figures it can be seen that the magnitudes of the MMGs are still high enough to be detected. The resulting relaying signals $|S_{1S}| < \delta$ and $|S_{2S}| > \delta$ are shown in Fig. 7.9c, in which a reverse direction to the fault is determined and a blocking signal is issued.

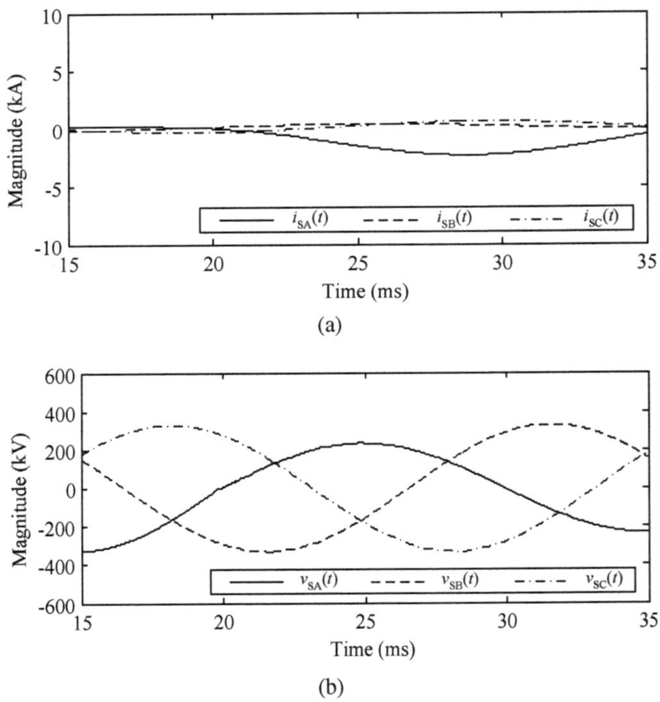

Fig. 7.8 The three-phase currents and voltages measured at busbar S under a phase-A-ground fault with a nearly 0° fault inception angle at 47 km from busbar S in section Q. **a** Three-phase currents measured at relaying position S. **b** Three-phase voltages measured at relaying position S

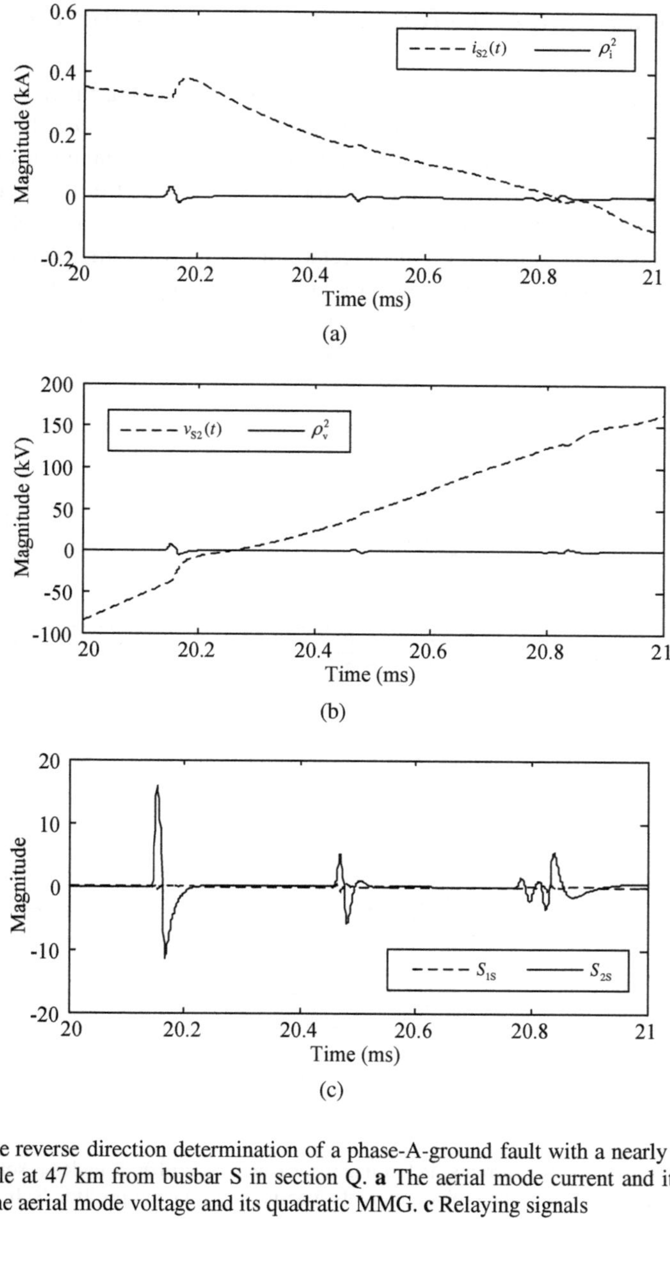

Fig. 7.9 The reverse direction determination of a phase-A-ground fault with a nearly 0° fault inception angle at 47 km from busbar S in section Q. **a** The aerial mode current and its quadratic MMG. **b** The aerial mode voltage and its quadratic MMG. **c** Relaying signals

7.3.4 A Solid Fault at the Busbar in the Reverse Direction

This case simulates a phase-A-ground fault taking place at busbar S, immediately behind the relaying position. The fault current and voltage signals are shown in Figs. 7.10a and b, respectively. The quadratic MMGs of transient aerial mode current and voltage are given in Figs. 7.11a and b, respectively. As noted in Fig. 7.11c, a disturbance occurs on the signal S_{1S} at the time of around 25 ms. This phenomenon is caused by the superimposition of reflected transients from the remote busbar R, which appears as the incident wave. Based on this limiting condition, a preset confirmation time T_{set} can be determined. Hence, within this time period, $|S_{1S}| < \delta$ and $|S_{2S}| > \delta$, a fault in the reverse direction is detected.

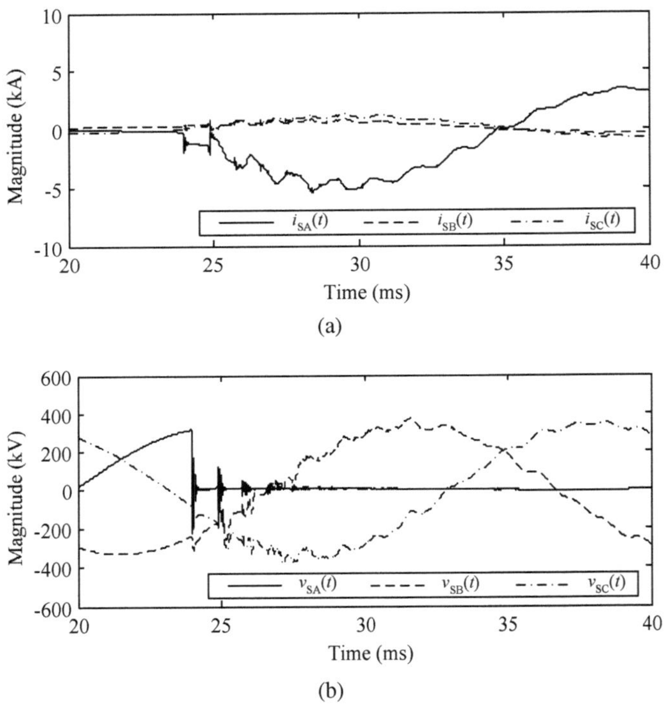

Fig. 7.10 The three-phase currents and voltages measured at busbar S under a solid phase-A-ground fault at busbar S. **a** Three-phase currents measured at relaying position S. **b** Three-phase voltages measured at relaying position S

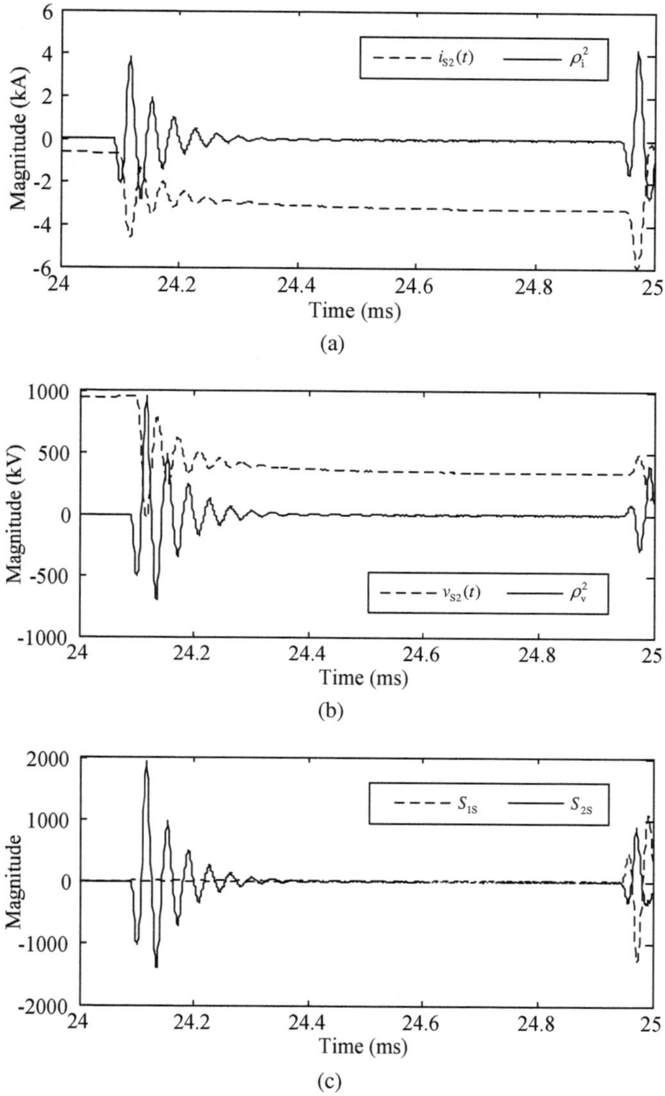

Fig. 7.11 The reverse direction determination of a solid phase-A-ground fault at busbar S. **a** The aerial mode current and its quadratic MMG. **b** The aerial mode voltage and its quadratic MMG. **c** Relaying signals

7.3.5 Faults of Different Types, Locations, Directions and Inception Angles

Apart from the four typical fault scenarios analysed previously, other types of fault have also been simulated with respect to different fault phases, locations, directions and inception angles. The extraction results of two of these are illustrated in Figs. 7.12–7.15, respectively. Figure 7.12 shows the three-phase current and voltage signals measured at busbar S under a phase-A-B fault at 100 km from busbar S in section P, and Figs. 7.13a and b show the MMGs of the transient aerial mode current and voltage, respectively. From the relaying signals, S_{1S} and S_{2S}, shown in Fig. 7.13c, we can see that they meet the criterion of $|S_{1S}| > \delta$ and $|S_{2S}| > \delta$, which reveals a forward fault. Another example is given in Figs. 7.14 and 7.15, in which a phase-A-B-ground fault occurs at 120 km from busbar S in section Q. From Fig. 7.15c it can be seen that $|S_{1S}| < \delta$ and $|S_{2S}| > \delta$. Therefore, it can be derived that there is a fault occurring in the reverse direction.

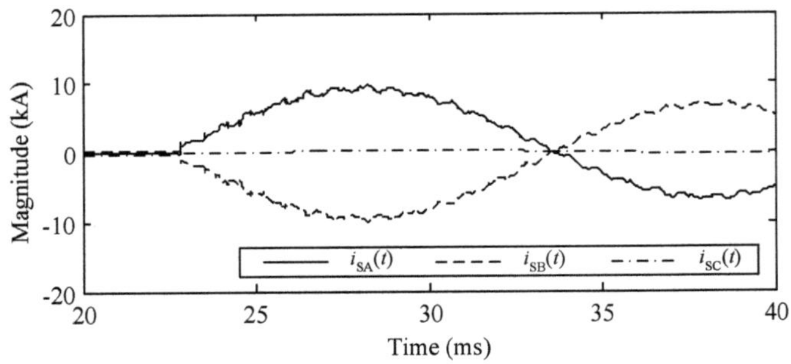

(a)

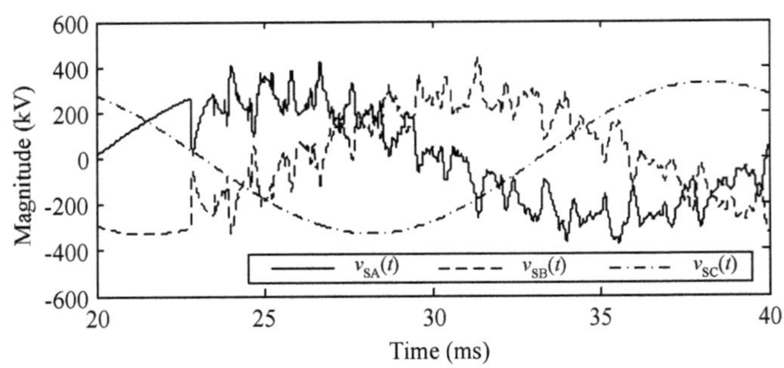

(b)

Fig. 7.12 The three-phase currents and voltages measured at busbar S under a phase-A-B fault at 100 km from busbar S in section P. **a** Three-phase currents measured at relaying position S. **b** Three-phase voltages measured at relaying position S

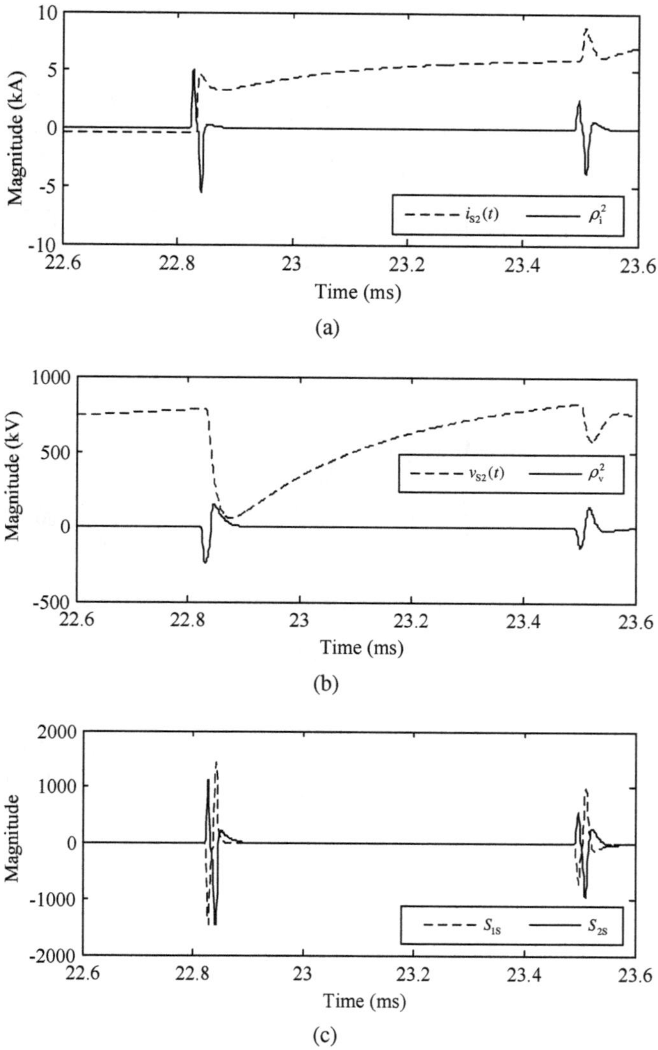

Fig. 7.13 The forward direction determination of a phase-A-B fault at 100 km from busbar S in section P. **a** The aerial mode current and its quadratic MMG. **b** The aerial mode voltage and its quadratic MMG. **c** Relaying signals

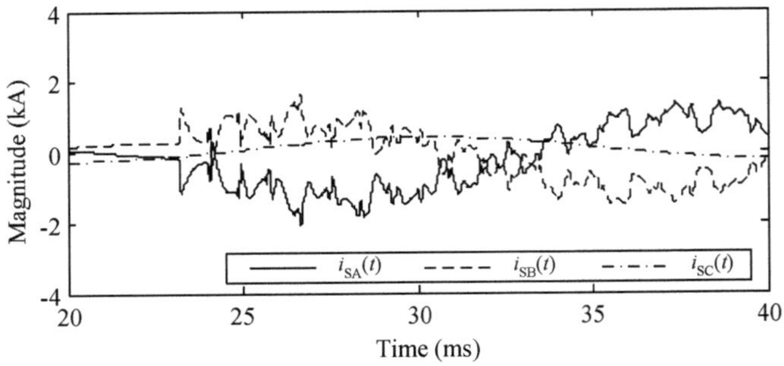

(a)

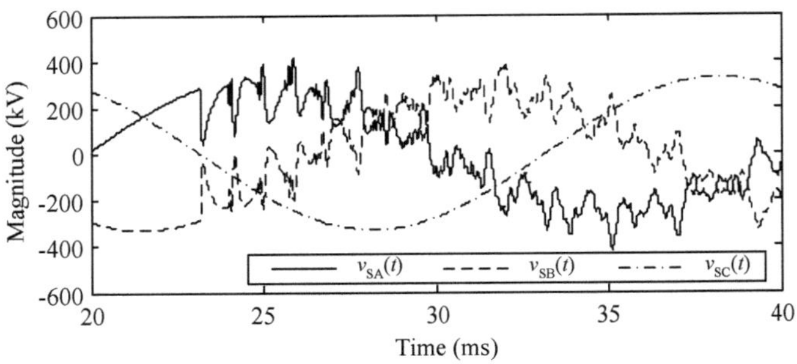

(b)

Fig. 7.14 The three-phase currents and voltages measured at busbar S under a phase-A-B-ground fault at 120 km from busbar S in section Q. **a** Three-phase currents measured at relaying position S. **b** Three-phase voltages measured at relaying position S

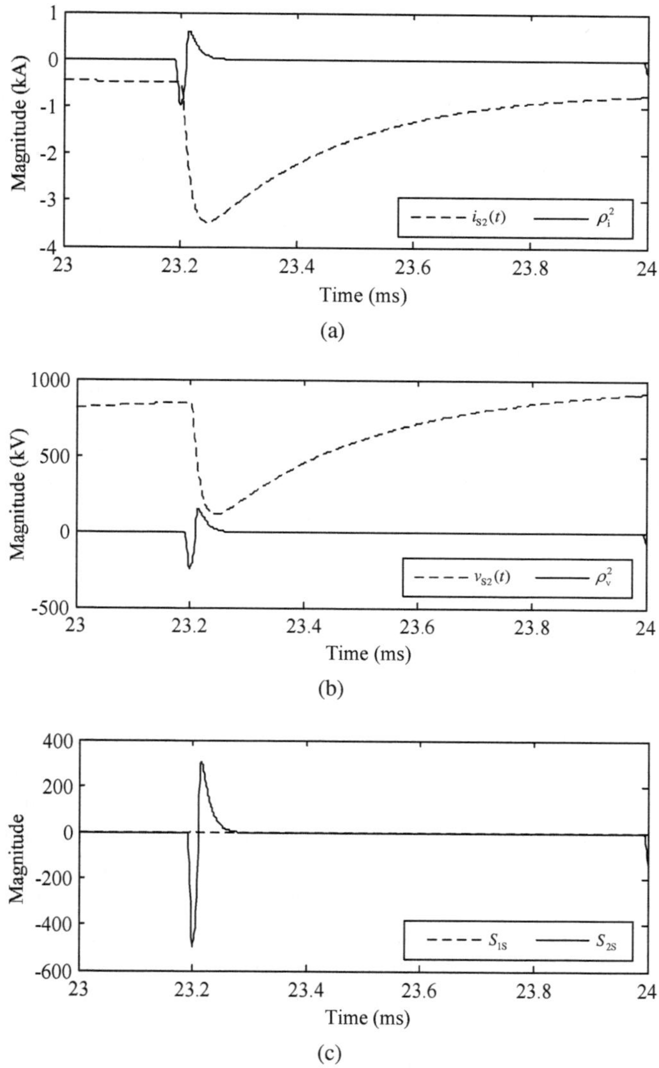

Fig. 7.15 The reverse direction determination of a phase-A-B-ground fault at 120 km from busbar S in section Q. **a** The aerial mode current and its quadratic MMG. **b** The aerial mode voltage and its quadratic MMG. **c** Relaying signals

7.3.6 Phase Selection with Different Fault Types

Three typical types of fault involving single-phase, double-phase and three-phase ones are evaluated and the results are shown in Figs. 7.16–7.18, respectively. All the faults are simulated to occur at 48 km from busbar S in section P. By examining the discriminants $|S_{\text{ph1}}|$ to $|S_{\text{ph6}}|$, all the simulation results can give the correct fault phase selections with reference to Table 7.1. For instance, among all the results shown in Fig. 7.16, only $|S_{\text{ph1}}| < \delta$, which indicates that the fault is a phase-A-ground one. The results of Fig. 7.17 show that except for $|S_{\text{ph4}}| < \delta$, all other discriminants are larger than the threshold. Hence, according to Table 7.1, the conclusion can be drawn that a phase-A-B fault occurs on the transmission line. This conclusion is in accordance with the simulation. As for the phase-A-B-C-ground fault, all its discriminants should be larger than threshold δ. The results shown in Fig. 7.18 satisfy this condition; therefore, a phase-A-B-C-ground is detected. These examples demonstrate the validity of the phase selection scheme. Nonetheless, the scheme has been tested on many other types of faults, and the results are as satisfactory as those presented here.

7.4 Summary

This chapter has presented an MM-based scheme for UHS protective relaying. The MMGA described in Sect. 2.4.4 is used to extract the wavefronts. MMGA can provide correct response to a transmission line fault under different fault types and locations. In contrast to FT and WT, where outputs indicate the integral value of the input signal components over a given interval, the computation of MMG is carried out within a short shifting data window. In the simulation studies, the data window of the MMGA requires eight samples only, which covers a period of 8 μs at the sampling rate of 1 MHz. The polarities of the transient wavefronts can be accurately identified by the positive or negative pulses in their MMGs at level 2, and the pulses have a fairly regular and explicit shape, which are easily recognised in terms of their maximum and polarity. Simulation results have shown that with this scheme, the reliability and sensitivity of UHS protection can be greatly improved.

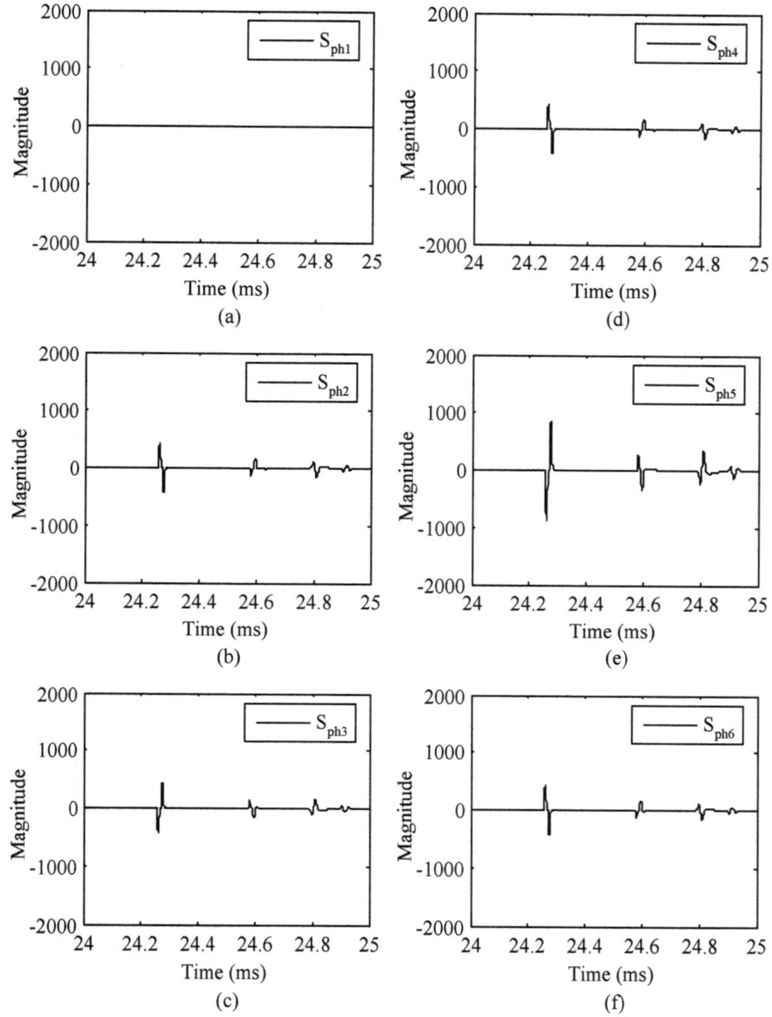

Fig. 7.16 The fault phase selection to a solid phase-A-ground fault at 48 km from busbar S in section P

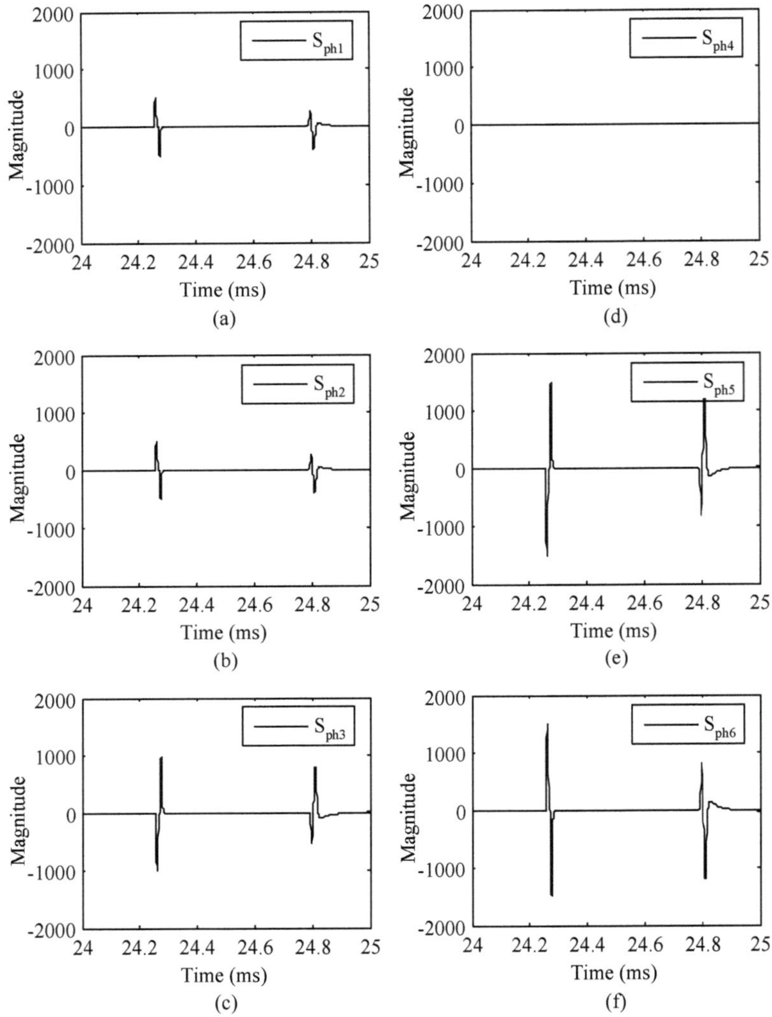

Fig. 7.17 The fault phase selection to a solid phase-A-B fault at 48 km from busbar S in section P

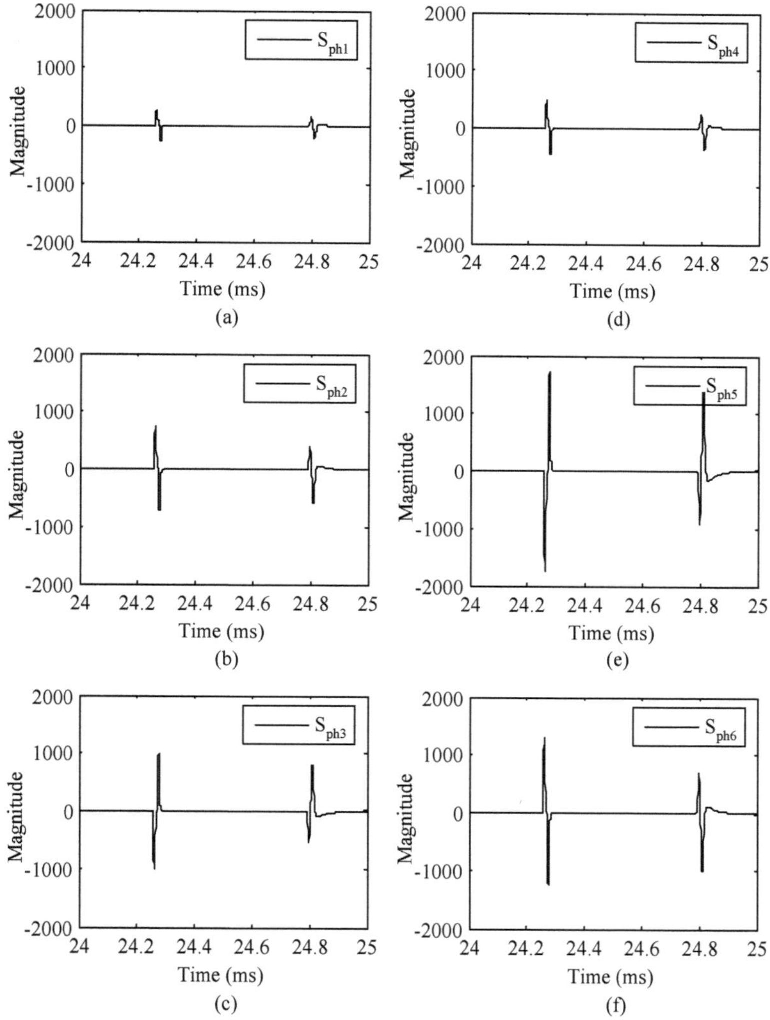

Fig. 7.18 The fault phase selection to a solid phase-A-B-C-ground fault at 48 km from busbar S in section P

Chapter 8
Fault Location on Transmission Lines

Abstract The content of this chapter can be viewed as the consequent processing of the directional protection discussed in the previous chapter. After deriving the direction discriminants of a fault taking place on a transmission line, the location of the fault can be determined by extracting the transient features. Fault location requires a more detailed and accurate information of the fault. Hence, it is essential to remove the noise that pollutes the transient waveforms. The morphological lifting scheme (MLS) discussed in Sect. 2.4.2 is used to remove the noise while preserving the gradient information of the wavefront, so that in the following processing, the wavefront can be correctly extracted by MMGA. Based on the polarities and the time-tags between the extracted wavefronts, the fault location or section can be identified. This chapter also introduces another approach that can deal with noise reduction and transient feature extraction at the same time. The approach is the so-called morphological undecimated wavelet (MUDW), which is a decomposition scheme designed originally for fault location, yet can be applied to other signal processing problems.

8.1 Introduction

Accurate fault location in a transmission line system is very desirable in order to enable electric utilities to quickly pinpoint the location of the disturbance [153, 154]. The application of these techniques effectively leads to a reduction of fault restoration times and the system operation cost during maintenance. Consequently, the reliability and quality of power delivery can be improved and the economic loss caused by line outages can be reduced.

Since transient-based protection techniques are based on the extraction of the characteristics of transient components generated by a fault and its concomitant arcing, a high sampling rate is always set to guarantee that the wavefronts of transients are caught. Consequently, the sampled signal of the transient may easily be disturbed by noise. If the noise cannot be eliminated, a transient-based protective relaying algorithm is not applicable, since the wavefronts are usually very faint and cannot be detected under such noisy conditions. Most of the existing tools used to extract the features of wavefronts are sensitive to noise and may enlarge the noise while extracting the gradient information of a wavefront. For example, a traditional low pass filter is used to restrain high-frequency noise to some extent. However, at the same time, the gradients of the wavefronts are attenuated as well.

In this chapter, MLS is presented to extract the wavefronts of transients for UHS protection relays in a noisy environment. MLS is a signal decomposition scheme to eliminate noise without affecting the shape of the transient. Afterwards, MMGA is employed to detect the characteristics of the transient voltage or current signals [145]. Consequently, the fault locations can be determined based on the detection results. The framework of first removing the noise, then extracting the wavefront ensures that the gradient information, which is essential for wavefront extraction, is highly preserved in the first stage.

Alternatively, the faults can be located by MUDW, a multi-resolution decomposition scheme developed based on morphological wavelets. For the introduction of multi-resolution decomposition schemes and morphological wavelets, readers are referred to Sect. 2.4 and [55] and [60]. The analysis operators and synthesis operator of MUDW are constructed according to the definition of morphological wavelets. Such a scheme, composed of morphological operators, totally inherits the simple computation property of the MM operation. The signal analysis operators of MUDW contain two parts, one is the morphological gradient operator that plays the same role as the MMGA and the other suppresses the noise. Such a construction is efficient for extracting features from a noisy transient signal, which is valuable for accurate fault location. Sect. 8.5 introduces the algorithm of MUDW and presents its performance on extracting the transient features directly from fault-generated signals, which may be contaminated by noise.

8.2 Principles of Fault Location

The transient fault locators are classified into types A, B, C and D according to operation modes [88]. Types A and D are passive methods relying on the high frequency transient components generated by the fault and its arcing. Type A is a single-ended technique in which transient sequences are captured at one end of the transmission line. Therefore, the distance to the fault is calculated based on transient propagation theory. The propagation theory of the fault-generated transients in a transmission line system is given in detail in Appendix A. Type D is a double-ended technique in which the fault distance is obtained based on measuring the time difference between the first fault-generated transients at opposite ends of a transmission line. Types B and C are active methods that use pulse generating circuitry. A new active method referred to as type E has been introduced in recent papers, which is similar to type A, but takes advantage of line energising transients generated by the reclosure of a circuit breaker (CB) when a permanent fault occurs.

8.2.1 Transient-Based Positional Protection

Transient positional protection is a type of transient-based protection widely used in EHV transmission lines. It can quickly and accurately pinpoint the fault location with information obtained from transients. Transient positional protection uses the arrival of successive transient wavefronts to identify the fault location at transmission lines.

A typical model of a transmission line system is shown in Fig. 8.1, which is a layout of 400 kV EHV transmission lines used in the U.K. SuperGrid system.

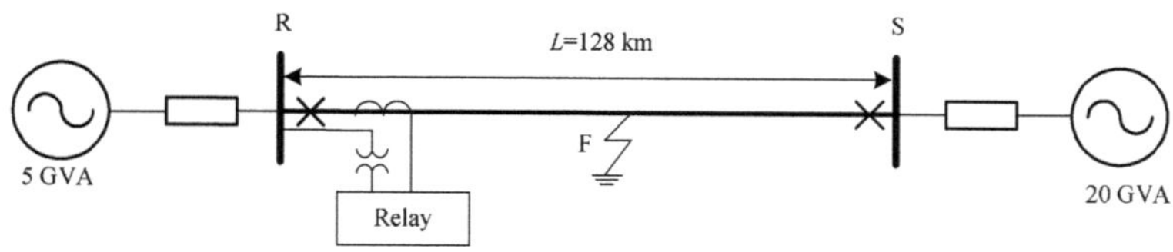

Fig. 8.1 A single-line transmission line model for simulation study

The principle of transient transmission is illustrated in Fig. 8.2. Once a fault occurs at F, a point in the first half of the transmission line m-n, the transient voltage or current signal travels along the transmission line to both directions until it reaches the busbars, R and S. The wave passed into the adjacent section and the rest is reflected backward. Let k_{m} and k_{n} be the reflection coefficients at ends m and n, respectively. Let K_{R} and K_{T} represent the reflection and refraction coefficients at the fault location, F. Suppose τ and τ' are the time periods as the first sequence of wavefronts arrive the busbars m and n, respectively. Therefore, the transients' voltages are represented as:

$$\begin{aligned} u_1(t) = &\ k_{\mathrm{m}} u_{\mathrm{F}}(t-\tau) + K_{\mathrm{R}} k_{\mathrm{m}}^2 u_{\mathrm{F}}(t-3\tau) \\ &+ k_{\mathrm{m}} k_{\mathrm{n}} K_{\mathrm{T}} u_{\mathrm{F}}(t-\tau-2\tau') + \cdots, \end{aligned} \tag{8.1}$$

$$\begin{aligned} u_2(t) = &\ u_{\mathrm{F}}(t-\tau) + K_{\mathrm{R}} k_{\mathrm{m}} u_{\mathrm{F}}(t-3\tau) \\ &+ k_{\mathrm{n}} K_{\mathrm{T}} u_{\mathrm{F}}(t-\tau-2\tau') + \cdots, \end{aligned} \tag{8.2}$$

where u_{F} is the voltage of the transient generated at point F; u_1 and u_2 are the voltages of the forward and reverse transients observed at end m, respectively. In transient-based protection, the transient voltages can be theoretically calculated using (8.1) and (8.2).

The sequence of wavefronts in transient voltages can be captured using the method given in this chapter. The wavefronts contain sufficient information on the transient fault, which can be used to locate the fault. Let x and y be the distances from point F to ends m and n, respectively. With the knowledge of the lengths of the transmission lines and the propagation speed of the high-frequency transient signal,

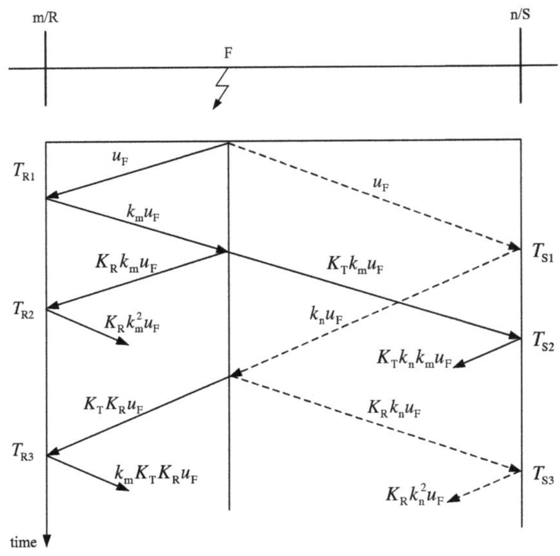

Fig. 8.2 Analysis of a typical transient grid

the fault location can be calculated and confirmed by the following equations:

$$\begin{aligned} 2\tau &= 2x/c, \\ 2\tau' &= 2y/c, \end{aligned} \tag{8.3}$$

where c is the propagation speed of high frequency signals on the line conductor.

8.2.2 Type A: A Passive Method Single-Ended Fault Locator

8.2.2.1 A Fault on the Second Half of the Line

Figure 8.3 shows a double-source transmission line system with a fault, which occurs on the second half of line RS, and its Bewley–lattice diagram. A transient current or voltage signal generated by the fault at point F travels away from F at the speed of c. It arrives some time later at busbars R and S, where part of the waveform passes into the adjacent section and the rest is reflected backwards. This process continues until the transient wavefronts become indistinct due to their multi-reflection and attenuation. The operating principle of a type A fault locator is developed on the successive identification of the high-frequency transients arriving at the measurement point. Reference to the first and subsequently captured transients, including their polarities, allows the distance to the fault from each end of the line, as shown in Fig. 8.3, to be calculated from the following formulae:

$$L_R = L - (T_{R2} - T_{R1})c/2, \tag{8.4}$$
$$L_S = (T_{S2} - T_{S1})c/2, \tag{8.5}$$

where L_R and L_S indicate the measured distance between the fault and busbars R and S, respectively; T_{R1}, T_{R2} and T_{S1}, T_{S2} are the times when the captured transient sequences are observed; c is the propagation speed on the transmission line; and L is the full length of the transmission line.

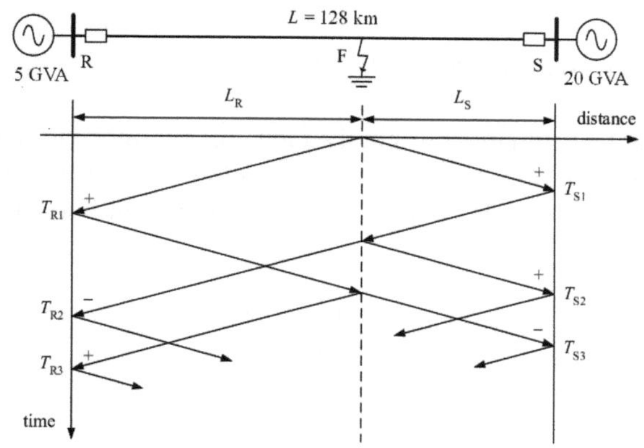

Fig. 8.3 Single-line diagram of a 400 kV power transmission line system and its Bewley–lattice diagram under a fault condition

In order to eliminate errors caused by the usage of a theoretical speed, an alternative calculation uses the first three transient sequences instead of the theoretical speed c. Formulae (8.4) and (8.5) are therefore modified as:

$$L_R = (T_{R3} - T_{R1})L/((T_{R2} - T_{R1}) + (T_{R3} - T_{R1})), \tag{8.6}$$
$$L_S = (T_{S2} - T_{S1})L/((T_{S2} - T_{S1}) + (T_{S3} - T_{S1})). \tag{8.7}$$

The validity of the above formulae depends mainly on the recognition of the first two wavefronts of the transient for (8.4) and (8.5) or the first three wavefronts for (8.6) and (8.7), in terms of their maxima and polarities. This is why the single-ended fault locator requires an effective signal processing technique to enable an explicit discrimination of the wavefronts. For conventional methods, it is difficult to achieve this goal because the features of the fault-generated transient wavefronts change very quickly. Furthermore, their shapes are affected by attenuation, multi-reflection and different fault conditions.

8.2.2.2 A Fault on the First Half of the Line

When a fault occurs on the first half of the transmission line, (8.4)–(8.7) should be modified. The determination of the fault location depends on the polarities of the second and third wavefronts, i.e. polarities $(+,+,-)$ and $(+,-,+)$ for the fault on the first and second half of the line at each measurement point, respectively. The positive and negative polarity can be pre-defined according to the polarity of the first wavefront captured. Sect. 8.4 will demonstrate how to define the reference polarity.

8.2.2.3 A Fault with a Low Fault Resistance

There is another fault situation, where the fault arc is assumed to remain ionised for an extended period of time and to present a very low resistance. From the reflection coefficient defined in (A.14) of Appendix A.3 at the point F, it can be obtained as follows:

$$K_R = \frac{Z_{0SF}^* - Z_{0RF}}{Z_{0SF}^* + Z_{0RF}},$$

with

$$Z_{0SF}^* = \frac{Z_{0SF} R_F}{Z_{0SF} + R_F},$$

where Z_{0SF} and Z_{0RF} are the equivalent surge impedances viewed from point F to busbars S and R, respectively, and R_F is the equivalent fault resistance. If $R_F = 0$, the reflection coefficient $K_R = -1$. Therefore, all the energy in any transients that arrives at the point is totally reflected with the same polarities as shown in Fig. 8.4.

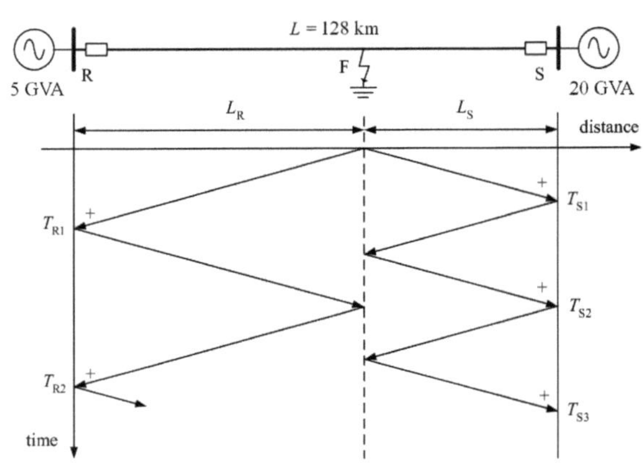

Fig. 8.4 Single-line diagram of a 400 kV power transmission line system and its Bewley–lattice diagram under a fault condition with a very low fault resistance

Such a situation can be determined by recognising the first three transient wavefronts with the same time interval and the same polarities as T_{S1}, T_{S2} and T_{S3} in Fig. 8.4. The fault distance can be calculated similarly using (8.5) at each end of the line.

8.2.3 Type D: A Passive Method Double-Ended Fault Locator

A type D fault locator operates with the first wavefront observed at each end of the line. The advantage of this method lies in avoiding the need to identify multi-reflecting transients. However, an accurate time reference system like GPS is necessary to synchronise the time-tagging at each end of the line. The distances from busbars R and S to the fault, as indicated in Fig. 8.3, are calculated according to the time difference between T_{R1} and T_{S1} by the following equations:

$$L_R = (L + (T_{R1} - T_{S1})c)/2, \tag{8.8}$$

$$L_S = (L + (T_{S1} - T_{R1})c)/2. \tag{8.9}$$

8.2.4 Type E: An Active Method Single-Ended Fault Locator

The technique of type E locator is similar to that of the type A locator, but takes advantage of line energising transients generated by the reclosure of a CB when a permanent fault occurs. When a CB re-closes with a permanent fault existing on the transmission line, the transient current or voltage signals initiate from the point at which the power source is injected to re-energise the line. Assuming the CT is installed behind the CB near busbar R, the first and subsequent transients observed at the measurement point are reflected either from the fault location, if the fault exists, or from the opposite end of the line, which could be either an open circuit if the opposite CB has opened, or the busbar. The propagation of the transient in the case of a permanent fault and opposite CB opening is illustrated in Fig. 8.5. In such a situation at the discontinuity point S, the reflection coefficient in (A.14) in Appendix A.3 can be obtained from:

$$K_R = \frac{Z_{0S} - Z_{0RS}}{Z_{0S} + Z_{0RS}},$$

where Z_{0RS} is the line surge impedance and $Z_{0S} = \infty$ is the sure impedance viewed from busbar S outwards. Hence, $K_R = 1$, and the incident transient is completely reflected with the same polarity at busbar R of the line, i.e. the polarities of the first wavefronts detected at busbar R are $(+,+,-)$ as indicated in Fig. 8.5. If the opposite CB remains closed when the line is energising, the reflection coefficient K_R will have a negative value, and the polarities will be $(+,+,+)$. Therefore, for a permanent fault, the type E fault locator can be used to reaffirm the fault distance in

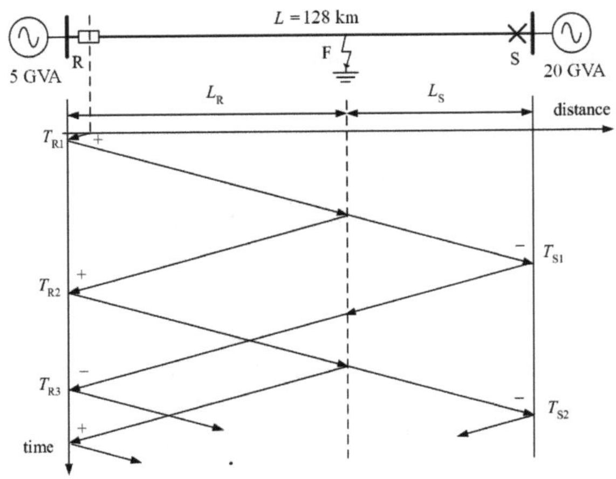

Fig. 8.5 Single-line diagram of a 400 kV power transmission line system and its Bewley–lattice diagram under re-closing under a permanent fault condition

support of types A and D. Equations (8.4)–(8.7) also apply to a type E fault locator to calculate the fault distance.

8.2.5 Online Measurement of the Transient Wave Propagation Speed

It has been discussed that, in practice, there is a slight difference between the measured and calculated values of the propagation speed of the transients. However, since the transients travel at a speed approximate to light, a tiny error in propagation speed will cause a considerable deviation in the measured fault distance. An alternative online measurement of the propagation speed is suggested to use the transients generated by line-energising, instead of using the calculated speed in the fault distance calculation. Figure 8.6 illustrates the propagation of the transients caused by the switching-in of CB at busbar R while the opposite CB is opening. The measured propagation speed can be obtained from:

$$c = 2L/(T_{R2} - T_{R1}). \tag{8.10}$$

It should be mentioned that the use of switching-in transients is a suggestive method. The following simulation will demonstrate its advantages, including active fault location and online measurement of the propagation speed. But in practice, such a technique is too complicated to implement, because the three poles of the CB can hardly contact simultaneously due to the inconsistence of the operating mechanism and contact arcing with different switching-in angles on transients. Therefore, such a method should be worthy of further investigation.

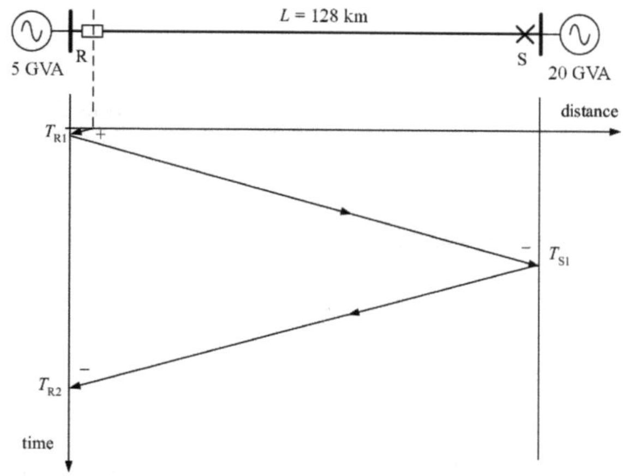

Fig. 8.6 Single-line diagram of a 400 kV power transmission line system and its Bewley–lattice diagram with line-energising

8.3 Noise Removal of Transients

The existence of noise greatly contaminates the transients due to the high sampling rate. As a result, the gradient of the wavefronts is quite small and it is usually difficult to detect the wavefronts. In order to guarantee accurate extraction of the wavefronts of the transients, the noise should be eliminated beforehand. Meanwhile, the shape information of the wavefronts should not be affected. This section presents an MLS for noise removal under the condition of shape preservation.

8.3.1 The Edge-Avoiding Prediction

As stated in Sect. 2.4.2, the first step of the lifting scheme is to split all samples into two categories: even samples and odd samples. An example is shown in Fig. 8.7, where '$\circ$' denotes odd samples $x_o[n]$ and '$\times$' denotes even samples $x_e[n]$, respectively. Also note that the figure contains the waveform of an ideal step edge.

As for the second step, prediction, a principal task is to choose a predictor window, based on the local properties of the transient. In order to preserve the gradient of a wavefront, samples belonging to the other step must not be included in the current predictor window, which means that the length of the prediction window should decrease automatically as it approaches the edge. A solution is used here to check whether the samples within the prediction window are well approximated by a low-order polynomial. If they are, then a higher-order predictor with a longer window is selected; if the samples do not meet the smoothness criterion, the sample responsible for the failure is determined and classified as an edge sample or a discontinuity coefficient. In this manner, as the predictor window moves towards the edge, the

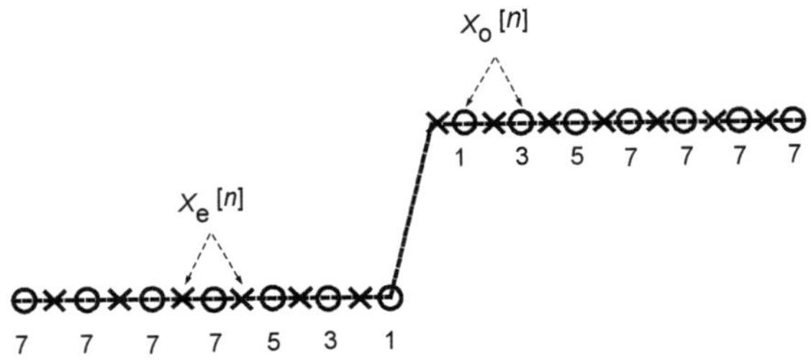

Fig. 8.7 An ideal step edge

order of the predictor is reduced so that the neighbourhood used for prediction will not overlap with the edge. Therefore, large errors are avoided around the edges in the lifting scheme.

The process of selecting these predictors near a step edge is illustrated in Fig. 8.7 as well, where a seventh-order predictor is used for non-edge samples, and the order of the predictor is reduced as the predictor window moves towards the edge.

8.3.2 Morphological Edge Reorganisation

A morphological operation is employed for edge reorganisation due to its ability to preserve or suppress the feature represented by SEs. The procedure of edge reorganisation is described as follows. Take the predictor window of seven samples, for example. As the centre of the predictor window moves to sample $s_{j,2l+1}$, the seven samples included in the window are:

$$[s_{j,2l-4},\ s_{j,2l-2},\ s_{j,2l},\ s_{j,2l+1},\ s_{j,2l+2},\ s_{j,2l+4},\ s_{j,2l+6}]. \tag{8.11}$$

To determine whether $s_{j,2l+6}$ is an edge sample, the gradients of the left samples and right samples of $s_{j,2l+4}$ are calculated and compared. For left samples, the morphological gradient is used, which is defined as the arithmetic difference between the results of dilation and erosion:

$$\rho(f,g) = f \oplus g - f \ominus g, \tag{8.12}$$

where f is the signal and g is the SE, and $\oplus$ and $\ominus$ represent dilation and erosion, respectively. On the other hand, for right samples, the absolute difference is used as the gradient, which is defined as:

$$\rho' = | f_{i+k} - f_i | . \tag{8.13}$$

In order to gain more accurate results, the comparison is undertaken several times and involves one more sample each time. For the left samples, a series of flat SEs,

g_i, with an increasing length, l_i ($l_i = i$, $i = 2,3,\ldots,6$), are employed to calculate the morphological gradient. The origin of g_i is the rightmost sample. In this example, it corresponds to sample $s_{j,2l+4}$. Hence, a group of morphological gradients are calculated as:

$$\begin{aligned} \rho_i &= \rho(s, g_i), \\ s &= \{s_{j,2l+6-2i}, \ldots, s_{j,2l+4}\},\ i = 2,3,\ldots,6. \end{aligned} \tag{8.14}$$

Another group of gradients is calculated for the right samples based on sample $s_{j,2l+4}$ as well:

$$\rho_i' = |s_{j,2l+2+2i} - s_{j,2l+4}|,\ i = 2,3,\ldots,6. \tag{8.15}$$

If $\forall\ i$, $\rho_i' \geq \rho_i$, then $s_{j,2l+6}$ is an edge sample. Therefore, the length of the predictor window is reduced to 5 and sample $s_{j,2l+6}$ should be left out.

8.3.3 Noise Filtering

The lifting scheme of a family of Deslauriers–Dubuc wavelets, as discussed in Sect. 2.4.2, is employed here to eliminate the noise. The families of wavelets have names in form of $(N, \tilde{N})$, where N is the number of vanishing moments of the analysing high-pass filter, while $\tilde{N}$ is the number of vanishing moments of the synthesising high-pass filter. In the prediction stage, a seventh-order predictor using the lifting scheme of $(6,2)$ Deslauriers–Dubuc wavelet is chosen for the region where no edge is detected. If an edge sample is detected, the order of the predictor window would be decreased to 5, 3 and 1 using the lifting scheme of $(4,2)$, $(2,2)$ Deslauriers–Dubuc and Harr wavelets, respectively. The lifting scheme of these wavelets is defined in Sect. 2.4.2.

The noise can be eliminated while the gradient of the edge is not affected and is kept in the prediction results. For instance, a step signal corrupted by noise is presented in Fig. 8.8a. For the purpose of comparison, a discrete wavelet transform (DWT), which also decomposes the signal into a certain approximation and a set of details, is applied to filter the noise. Figure 8.8b shows the de-noised approximation of DWT. Large errors are found at the samples around the edge. However, the shape of the step is well kept by MLS, as shown in Fig. 8.8c, due to the merit that the predictor window of MLS never overlaps the edge.

After the noise is eliminated, MMGA is used to extract the sequence of wavefronts of the transients. A protection scheme is designed for positional protection, as explicated in the following sections. Nonetheless, the scheme is also suitable for directional protection.

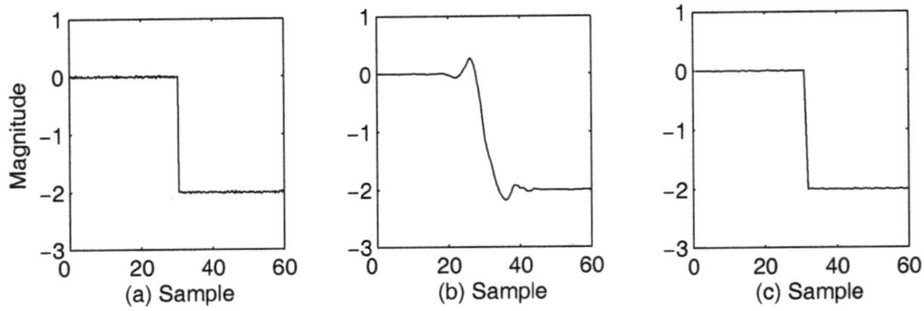

Fig. 8.8 **a** A step signal contaminated by noise. **b** The filtering result of DWT. **c** The filtering result of MLS

8.4 Accurate Fault Location by Morphological Filters

Simulation studies are carried out on the typical model of a transmission line system shown in Fig. 8.1. A variety of fault scenarios, including different fault inception angles, fault resistances and fault types, have been simulated to evaluate the validity of this approach. The sampling frequency in the simulation is 1 MHz. The currents and voltages mentioned in the following simulation studies are transformed by CT and VT, respectively. The MMGA used is a quadratic one (level $a = 2$) with a flat SE of length 5. With reference to the maxima and their polarities derived from the quadratic MMG, the distance to the fault can be obtained from the formulae given previously with respect to types A, D and E [153].

8.4.1 A Solid Phase-A-Ground Fault

Figure 8.9 shows a transient generated by a solid phase-A-ground fault occurring at a distance of 80 km from busbar R indicated in Fig. 8.1. The observed currents at busbar R are firstly transformed into aerial modal signals. It is noted that the fault-generated transient is superimposed on the power fundamental frequency component. For simulation purpose, the current transients observed at the ends of the transmission line are polluted by Gaussian white noise. The mean and variance of the noise in the simulation studies are 0 and 100 A, respectively.

MLS can eliminate most the noise without affecting the gradients of the wavefronts. For the purpose of illustration, the first wavefront indicated in Fig. 8.9 is enlarged and then shown in Fig. 8.10a. Figure 8.10b is the filtering result of MLS. In this figure, the noise is reduced and the magnitude of the wavefront is not affected. Based on the filtered signal, MMGA is used to extract the wavefronts. The peaks in Fig. 8.11 can be recognised as the locations of the wavefronts and the time-tags among the wavefronts are hereby obtained. The first two wavefronts in Fig. 8.11 have opposite polarities, which indicates that the second wavefront is reflected

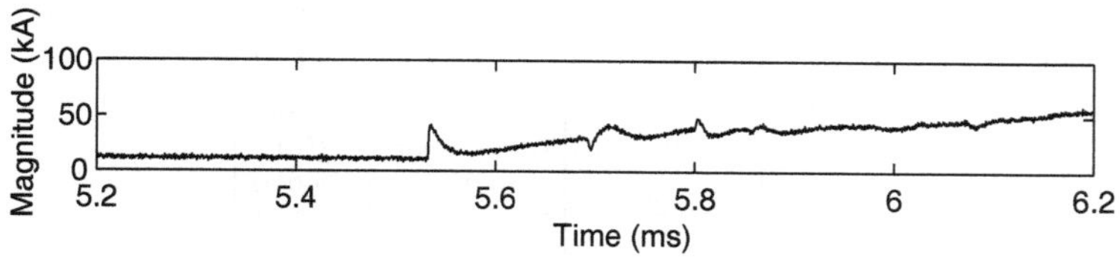

Fig. 8.9 A noise-polluted transient generated by a single-phase-to-ground fault at 80 km from busbar R

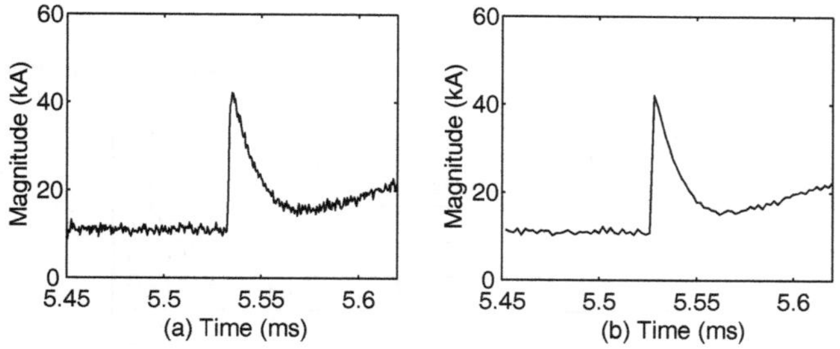

Fig. 8.10 **a** Enlargement of the first wavefront indicated in Fig. 8.9. **b** The filtering result of MLS

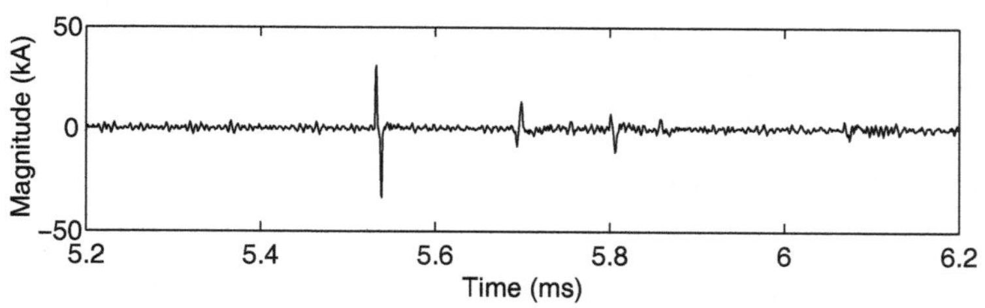

Fig. 8.11 Result of wavefront detection using MMGA (a phase-A-ground fault at 80 km from busbar R)

from the opposite busbar, S. Therefore, it can be concluded that there is a fault occurring at a location belonging to the second half of the transmission line. Then the distance to the fault location can be accurately calculated. The calculation process and equations are given in Sect. 8.2.1.

Simulation studies have been carried out under a variety of conditions of fault resistance and fault inception angles. The results are listed in Tables 8.1 and 8.2, respectively.

Table 8.1 The fault location results and errors (%) affected by factors of fault resistance (phase-A-ground fault)

Resistance (Ω)	Actual location (km)	Calculated location (km)	Error (%)
100	20	19.7321	0.2092%
	68	67.8098	0.1486%
	126	126.4668	0.3649%
200	20	19.7321	0.2092%
	68	67.8098	0.1486%
	126	126.4668	0.3649%

Table 8.2 The fault location results and errors (%) affected by factors of fault inception angle (phase-A-ground fault)

Inception angle (degrees)	Actual location (km)	Calculated location (km)	Error (%)
15°	20	19.7321	0.2092%
	80	80.1367	0.1067%
	108	107.7586	0.1886%
45°	20	19.7321	0.2092%
	80	80.1367	0.1067%
	108	107.7586	0.1886%
90°	20	19.7321	0.2092%
	80	80.1367	0.1067%
	108	107.7586	0.1886%

8.4.2 A Double-Phase-to-Ground Fault

In this case, a double-phase-to-ground fault (phase-A-B-ground) occurs at 20 km from busbar R in Fig. 8.1. The transient observed at busbar R is shown in Fig. 8.12. To illustrate the effectiveness of MLS, three wavefronts indicated in Fig. 8.12 are enlarged and shown in Fig. 8.13a. The wavefronts in Fig. 8.13b are more distinct than those indicated in Fig. 8.13a, due to the application of MLS. After the noise is eliminated by MLS, the wavefronts can be extracted by MMGA. The detection result of the wavefronts indicates that the fault involves two phases, and accurate response can be given based on the results indicated in Fig. 8.14. Though the magnitudes of wavefronts are reduced due to the ground-fault resistance, these indistinct wavefronts still can be recognised with the noise previously eliminated by MLS.

Simulation studies have also been carried out with a variety of different fault locations and the results are given in Table 8.3. In Sects. 8.4.1 and 8.4.2, a variety of tests were carried out with the following variations in the simulation condition:

- *Fault types:* phase-A-ground, phase-B-C, phase-A-B-ground, three-phase fault. The results are the same for different fault types.
- *Fault resistance:* 100 and 200 Ω (see Table 8.1).
- *Fault inception angle:* $\delta = 15^\circ, 45^\circ$ and 90° (see Table 8.2).
- *Fault locations:* Faults are introduced at 20, 68, 80, 108 and 126 km from busbar R (see Table 8.3).

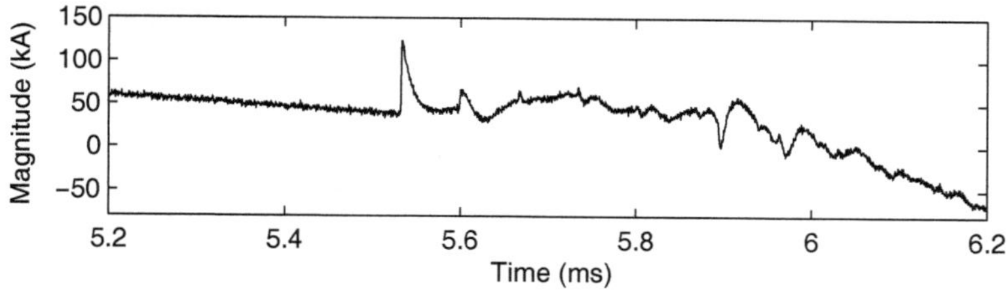

Fig. 8.12 A noise-polluted transient generated by a double-phase-to-ground fault at 20 km from busbar R

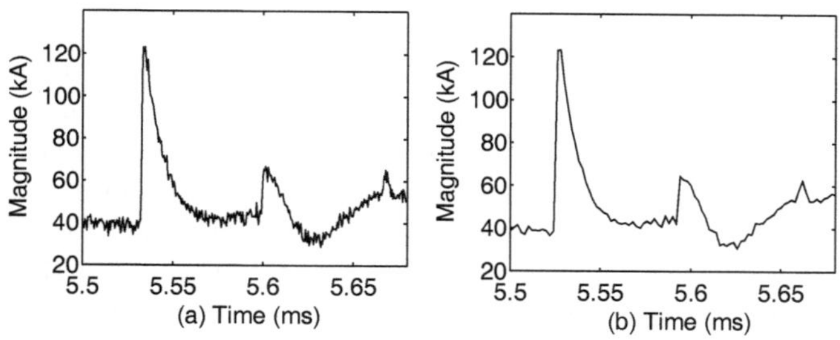

Fig. 8.13 **a** Enlargement of the wavefronts indicated in Fig. 8.12. **b** The filtering result of MLS

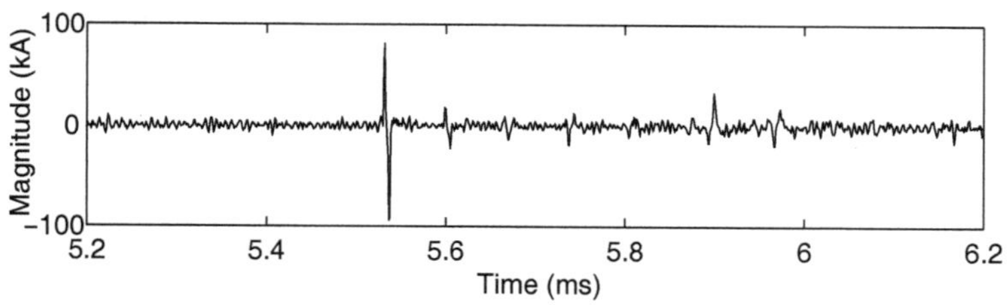

Fig. 8.14 Result of wavefront detection using MMGA (a phase-A-B-ground fault at 20 km from busbar R)

The results have shown that the accuracy of this scheme for fault location is not affected by the type, resistance, inception angle or the distance of the fault.

For the cases of fault inception at zero crossing, theoretically, no transient is generated from the fault location. Therefore, the transient-based protection relay cannot operate in this situation. A hybrid algorithm was introduced recently to overcome this theoretical shortcoming of transient-based protection, by including a traditional distance protective relaying algorithm into the transient-based protection relay [106].

Table 8.3 The fault location results and errors (%) affected by factors of fault locations (phase-A-B-ground fault)

Actual location (km)	Calculated location (km)	Error (%)
20	19.7321	0.2092%
68	67.8098	0.1486%
80	80.1367	0.1067%
108	107.7586	0.1886%
126	126.4668	0.3649%

8.4.3 A Fault Close to the Busbar

Having demonstrated the effectiveness of MLS, in the following studies, MMGA is carried out on noise-free signals. Generally, it is difficult to accurately detect the short distance to a fault that takes place very close to a busbar, because of multi-reflection and the high propagating speed of the transients. However, the quadratic MMGs, as shown in Fig. 8.15, are immune to these conditions, due to the short data window and the edge detection technique. Theoretically, a quadratic MMG with an SE of five samples is able to detect a fault occurring at a distance as close as 1 km to the busbar. The results attained for the fault location of a phase-A-ground fault at 2 km from busbar R are listed in Table 8.4. Three types of fault locators have been employed here to evaluate their performance. Due to the elimination of the speed, as presented by (8.6) and (8.7), the modified type A locator is much more accurate than the others.

Table 8.4 The fault location results of a phase-A-ground fault at 2 km from busbar R

Locator	End	Formula	Result (km)	Error (m)
Type A	R	(8.5)	2.064	+63.8
	S	(8.4)	126.084	+83.6
Modified type A	R	(8.7)	2.089	+88.6
	S	(8.6)	126.061	+60.6
Type D	R	(8.8)	2.234	+234.3
	S	(8.9)	125.766	-234.3

8.4.4 A Fault with a Stable Low Resistance

In order to satisfy a zero fault resistance, a solid phase-A-B fault at 100 km from busbar R is simulated. As expected, the fault-generated transient is completely reflected from point F. The wavefronts of the transient represented by the MMGs arrive at each end of the line with the same time interval and the same polarities, as

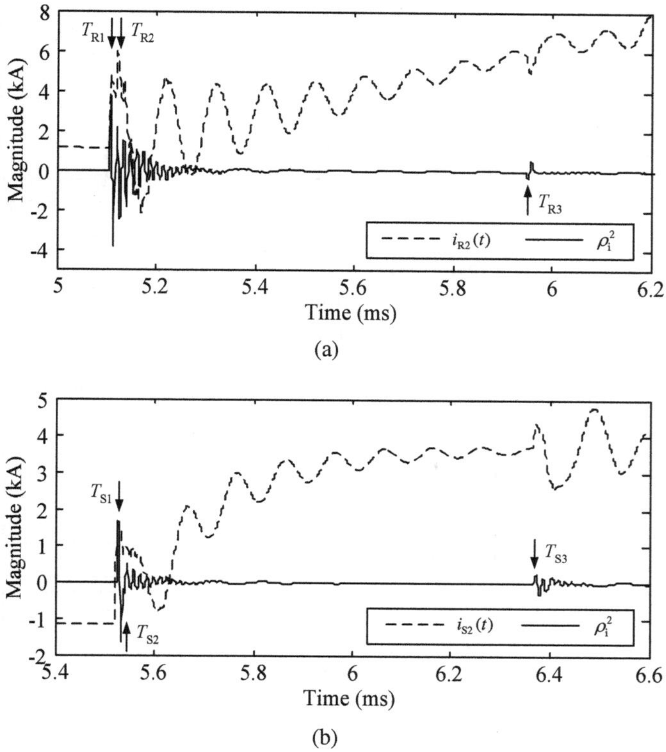

Fig. 8.15 The result of wavefront detection using MMGA (a phase-A-ground fault at 2 km from busbar R). **a** The aerial mode current at busbar R and its quadratic MMG. **b** The aerial mode current at busbar S and its quadratic MMG

shown in Fig. 8.16. The fault location is calculated by (8.5), (8.8) and (8.9), respectively, and the results are listed in Table 8.5.

Table 8.5 The fault location results of a phase-A-B fault at 100 km from busbar R

Locator	End	Formula	Result (km)	Error (m)
Type A	R	(8.5)	100.103	+102.6
	S	(8.5)	28.047	+46.6
Type D	R	(8.8)	99.953	-46.6
	S	(8.9)	28.047	+46.6

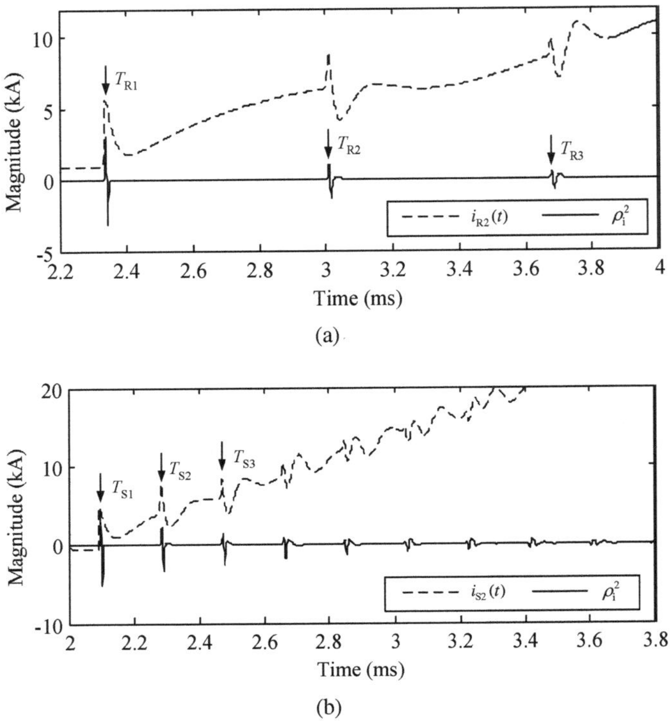

Fig. 8.16 The result of wavefront detection using MMGA (a phase-A-B fault at 100 km from busbar R). **a** The aerial mode current at busbar R and its quadratic MMG. **b** The aerial mode current at busbar S and its quadratic MMG

8.4.5 Switching-In Transients

When energising a transmission line with an existing permanent fault, the basic operating principle of type E fault locator is the same as that of type A, except that the first transient observed at busbar R is reflected from the fault point rather than originated from it. Figure 8.17 shows the modal current observed at busbar R and the result of the wavefront extraction. The results of fault location by a type E locator are shown in Table 8.6.

Table 8.6 The fault location results of a permanent phase-A-ground fault at 80 km from busbar R

Locator	End	Formula	Result (km)	Error (m)
Type E	R	(8.4)	80.533	+533.3
Modified type E	R	(8.6)	79.963	-37.3

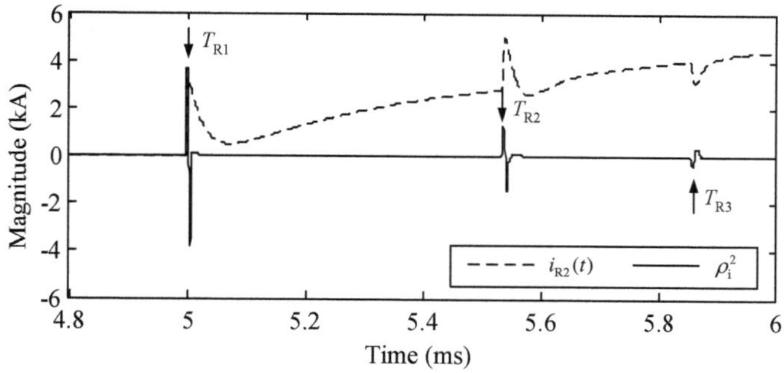

Fig. 8.17 The result of wavefront detection using MMGA for a type E locator (a permanent phase-A-ground fault at 80 km from busbar R)

8.4.6 Calculated Transient Wave Propagation Speed and Fault Locations

A transmission line energised by the CB at busbar R, the resulting transient aerial mode current observed at measurement point R and its MMG are shown in Fig. 8.18. The propagation speed measured online can be calculated as $c = 298,368,298$ m/s by applying (8.10). Compared with the value of the theoretical speed, there exists a slight deviation. By using the measured speed, the fault distance for a solid phase-A-ground fault demonstrated in Sect. 8.4.1 is recalculated; the results are listed in Table 8.7. It can be seen that the accuracy of the fault locators has been improved.

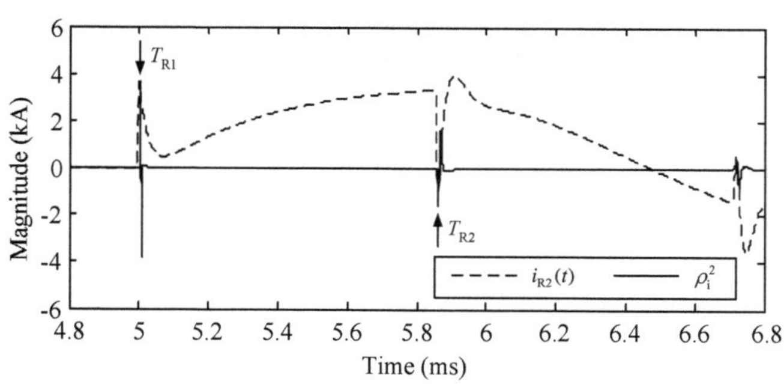

Fig. 8.18 The transient aerial mode current generated by line-energising and its MMG

Table 8.7 The fault location results of a phase-A-ground fault at 80 km from busbar R using online measured propagation speed

Locator	End	Formula	Result (km)	Error (m)
Type A	R	(8.4)	79.963	-37.3
	S	(8.5)	48.037	+37.3
Type D	R	(8.8)	79.963	-37.3
	S	(8.9)	48.037	+37.3

8.5 Morphological Undecimated Wavelet Decomposition for Fault Location

In this section, we present the decomposition scheme of MUDW for fault location on power transmission lines [154]. MUDW is developed based on morphological wavelets, as described in Sect. 2.4.3, for both noise reduction and feature extraction in signal processing. The analysis operators and the synthesis operator of MUDW strictly satisfy the perfect reconstruction condition [55]. In the application to fault location, MUDW is used to extract features from noise-imposed fault-generated transient voltage or current signals on power transmission lines.

8.5.1 Morphological Undecimated Wavelets

MUDW is developed based on morphological wavelets. It adopts not only the framework as shown in Fig. 2.8, but also the undecimated algorithm. The undecimated algorithm is based on the idea of no decimation. It applies the wavelet transform and omits both down-sampling in the forward transform and up-sampling in the inverse transform. More precisely, it applies the transform at each sample of the signal [87]. In signal processing, this algorithm may give the best result in terms of high quality filtering, by avoiding information loss and with less distortion, which may be caused by noise removal.

The analysis and synthesis operators are constructed using dilation (symbolised by δ) and erosion (symbolised by ε), as well as opening and closing (denoted by γ and ϕ, respectively). The signal analysis operator $\psi_j^{\uparrow}$ and the detail analysis operator $\omega_j^{\uparrow}$ of MUDW are defined by $\frac{1}{2}(\delta\phi - \varepsilon\gamma + \delta - \varepsilon)$ and $\mathsf{id} - \frac{1}{2}(\delta\phi - \varepsilon\gamma + \delta - \varepsilon)$, respectively. Here, id represents the identity transform: $\forall f$, $\mathsf{id}(f) = f$. Hence, we have:

$$x_{j+1} = \psi_j^{\uparrow}(x_j) = \frac{1}{2}(\delta\phi - \varepsilon\gamma + \delta - \varepsilon)(x_j), \tag{8.16}$$

$$y_{j+1} = \omega_j^{\uparrow}(x_j) = \left[\mathsf{id} - \frac{1}{2}(\delta\phi - \varepsilon\gamma + \delta - \varepsilon)\right](x_j). \tag{8.17}$$

Taking the synthesis operator $\Psi_j^{\downarrow}$ as algebra addition:

$$
\begin{aligned}
&\Psi_j^{\downarrow}(\psi_j^{\uparrow}(x_j), \omega_j^{\uparrow}(x_j)) \\
&= \frac{1}{2}(\delta\phi - \varepsilon\gamma + \delta - \varepsilon)(x_j) + \left[\mathrm{id} - \frac{1}{2}(\delta\phi - \varepsilon\gamma + \delta - \varepsilon)\right](x_j) \\
&= x_j,
\end{aligned} \tag{8.18}
$$

it is easy to obtain

$$\psi_j^{\uparrow}(\Psi_j^{\downarrow}(x_{j+1}, y_{j+1})) = \psi_j^{\uparrow}(x_j) = x_{j+1}, \tag{8.19}$$

$$\omega_j^{\uparrow}(\Psi_j^{\downarrow}(x_{j+1}, y_{j+1})) = y_{j+1}, \tag{8.20}$$

where $x_j \in \mathrm{V}_j$, $x_{j+1} \in \mathrm{V}_{j+1}$ and $y_{j+1} \in \mathrm{W}_{j+1}$. Therefore, the analysis operators and synthesis operator of MUDW satisfy the perfect reconstruction condition of (2.49) and (2.50). Furthermore, it is not difficult to prove that (2.51)–(2.53) are also satisfied. Thus, MUDW composed of (8.16) as the signal analysis operator, (8.17) as the detail analysis operator and (8.18) as the synthesis operator is a morphological wavelet scheme.

Since dilation after erosion constructs opening ($\gamma = \varepsilon\delta$) and erosion after dilation constructs closing ($\phi = \delta\varepsilon$), and $\varepsilon\delta\varepsilon = \varepsilon$, $\delta\varepsilon\delta = \delta$, we obtain

$$\gamma\varepsilon = \varepsilon\phi = \varepsilon, \tag{8.21}$$

$$\phi\delta = \delta\gamma = \delta. \tag{8.22}$$

Consequently, the signal analysis operator can be transformed to

$$
\begin{aligned}
\psi_j^{\uparrow} &= \frac{1}{2}(\delta\delta\varepsilon - \varepsilon\varepsilon\delta + \delta\varepsilon\delta - \varepsilon\delta\varepsilon) \\
&= \frac{1}{2}[(\delta - \varepsilon)\delta\varepsilon + (\delta - \varepsilon)\varepsilon\delta] \\
&= \frac{1}{2}(\delta\varepsilon + \varepsilon\delta)(\delta - \varepsilon), \quad j = 0, 1, 2, \ldots
\end{aligned} \tag{8.23}
$$

This equation shows that the signal analysis operator is composed of two parts: $\frac{1}{2}(\delta\varepsilon + \varepsilon\delta)$ and $(\delta - \varepsilon)$. The latter part is a morphological gradient operator that plays an important role in MMGA, while the function of the remaining part is to suppress noise.

8.5.2 Fault Location Using MUDW

The performance of MUDW on fault location is evaluated in this section. The simulation environment is the same as the one used in the previous simulation studies in this chapter.

8.5.2.1 Different Levels of MUDW for Fault Transient Extraction

Based on the expression of the analysis operator of MUDW (8.23) and its construction as shown in Fig. 2.8, the output of each level can be expressed as:

$$r_n = \frac{1}{2^n}(\delta\varepsilon + \varepsilon\delta)^n(\delta - \varepsilon)^n[i_{R2}(t)], \tag{8.24}$$

where r_n denotes the output on the nth level. Define $\psi_n^{\uparrow} = \frac{1}{2^n}(\delta\varepsilon + \varepsilon\delta)^n(\delta - \varepsilon)^n$, then when $n = 2$,

$$r_2 = \frac{1}{4}(\delta\varepsilon + \varepsilon\delta)^2(\delta - \varepsilon)^2[i_{R2}(t)]. \tag{8.25}$$

The operator $(\delta - \varepsilon)^2$ is the quadratic morphological gradient.

The second level of MUDW with a flat SE of length 5 is used in the simulation studies to process the transient voltage or current signals, and it is supposed that the wavefronts of the transients should be extracted from the output, $\psi_2^{\uparrow}$. With reference to the maxima derived from the $\psi_2^{\uparrow}$ of the aerial modal signals, the distance to point F can be obtained by (8.6) and (8.7) with respect to a type A fault locator.

Figure 8.19 illustrates the outputs of $\psi_2^{\uparrow}$, $\psi_3^{\uparrow}$ and $\psi_4^{\uparrow}$, when the signal-to-noise ratio SNR $= 0$. It is shown that with increasing n, the transient features extracted by MUDW are badly deformed. When $n = 3$ or $n = 4$, the polarities of the transient wavefronts cannot be determined. Not only the polarities are affected by the extraction, but also the accuracy of the determination deteriorates. Table 8.8 lists the errors of MUDW at different levels. Based on the trial results, MUDW at the second level, $\psi_2^{\uparrow}$, is chosen as the operator to fulfil the task of noise reduction and wavefront extraction.

Table 8.8 The fault location results and errors affected by the level of MUDW in a noise-free environment (phase-A-ground fault at 80 km from busbar R)

Level	Result (km)	Error (m)
$\psi_2^{\uparrow}$	80.056	+56
$\psi_3^{\uparrow}$	78.642	−1358
$\psi_4^{\uparrow}$	81.150	+1150
$\psi_5^{\uparrow}$	74.970	−5030

8.5.2.2 The Performance of MUDW on Wavefront Extraction

In order to demonstrate the robustness of MUDW in a noisy environment, a variety of noise with different SNRs has been artificially added to the input signal. Figures 8.20 and 8.21 illustrate the performance of MUDW with different noise disturbances and the determination of the transient features, respectively. Table 8.9 shows the results and errors of fault location of a phase-A-ground fault at 80 km from busbar

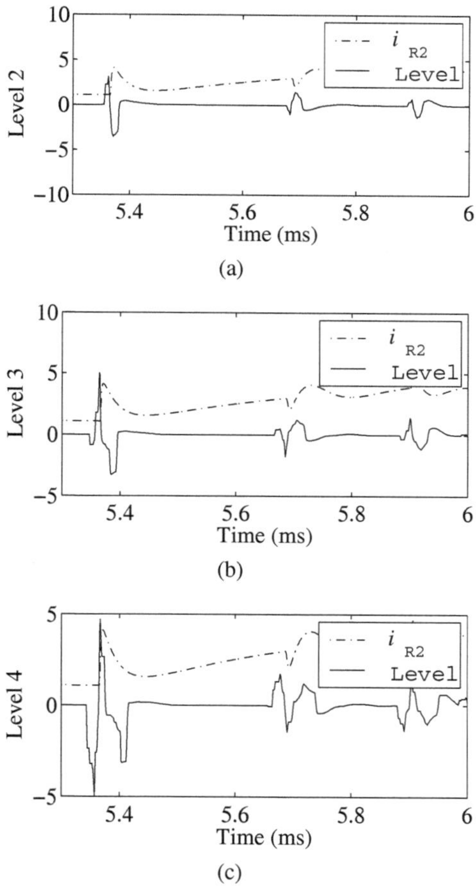

Fig. 8.19 The outputs on different levels of MUDW. **a** The second level output of MUDW. **b** The third level output of MUDW. **c** The fourth level output of MUDW

R, when the SNR varies from 45.77 dB to 19.01 dB. It is shown that the first time-tag T_{R1} can be extracted in almost all the circumstances when SNR varies from 45.77 dB to 19.01 dB. The second time-tag T_{R2} and T_{R3} can only be extracted when SNR is larger than 20.73 dB. A rough conclusion can be drawn from Table 8.9 that the limitation of MUDW to obtain an accurate result is an SNR of 25.80 dB. If SNR is less than this threshold, the accuracy of MUDW will be greatly damaged and the scheme becomes useless. Furthermore, as shown in the case of SNR= 30.28 dB, the error reaches 299 m. Although it is larger than the error when SNR= 25.80 dB, it is still within the requirement of the system, which is ±300 m [153].

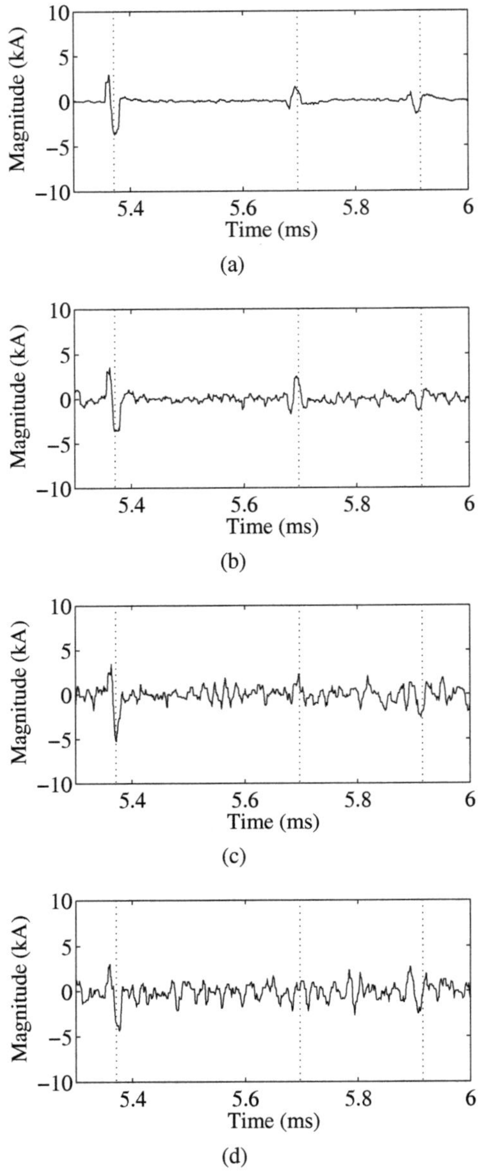

Fig. 8.20 The performance of MUDW under a variety of different noise conditions. The *dotted lines* illustrate the positions of time-tags T_{R1}, T_{R2} and T_{R3} when no noise is imposed on the signal. **a** SNR=39.82 dB. **b** SNR=25.80 dB. **c** SNR=20.73 dB. **d** SNR=19.01 dB

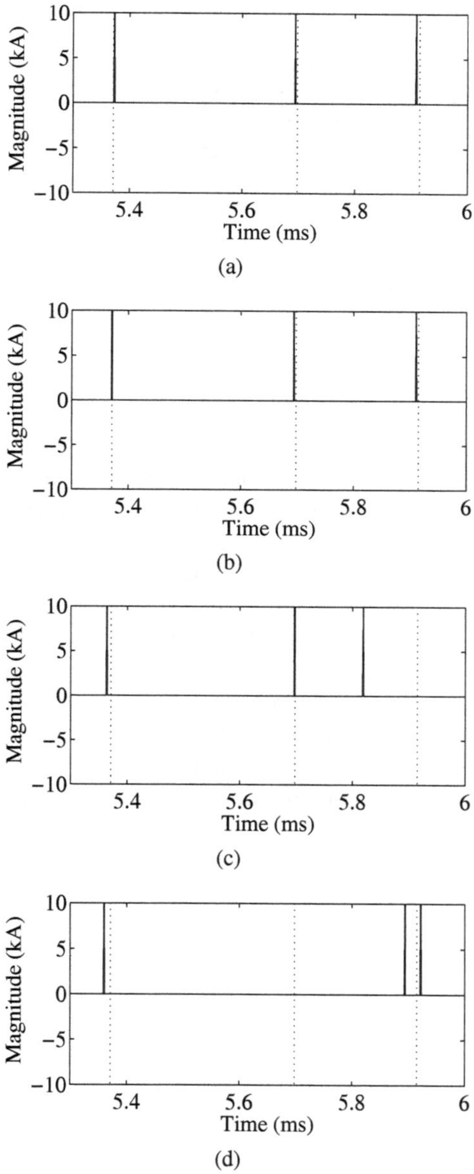

Fig. 8.21 The determination of the transient features after MUDW has extracted the noise. The *dotted lines* illustrate the positions of time-tags T_{R1}, T_{R2} and T_{R3} when no noise is imposed on the signal. **a** SNR=39.82 dB. **b** SNR=25.80 dB. **c** SNR=20.73 dB. **d** SNR=19.01 dB

Table 8.9 The fault location results and errors affected by SNR (phase-A-ground fault at 80 km from busbar R)

SNR (dB)	Result (km)	Error (m)
45.77	80.056	+56
39.82	80.074	+74
30.28	80.299	+299
25.80	80.093	+93
22.94	78.800	−1200
20.73	73.909	−6091
19.01	65.640	−14360

8.6 Summary

In this chapter, we have discussed the positional protection for UHS protective relaying. The positional protection aims to obtain the exact location of the fault occurring on the transmission line. Therefore, it is more complex than directional protection and requires a more accurate detection of the transient features of the voltage or current signals. Two algorithms for positional protection have been studied here. For the first one, the fault-generated transients on the transmission line are first denoised by MLS, so that the gradients of the wavefronts can be preserved while the noise is being removed. Afterwards, MMGA is used for the detection of the wavefronts. The second algorithm, MUDW, combines the undecimated algorithm with morphological wavelet, so that noise depression and wavefront recognition are performed at the same time. It has been demonstrated by the simulation results that both algorithms are able to accurately locate the fault in a noisy environment.

Appendix A
Electromagnetic Transient Analysis of Transmission Lines

A.1 Distributed Parameter Model of Transmission Lines

In order to portray transient phenomena with higher accuracy, a transmission line is treated as a distributed parameter circuit. The circuit presentation of the distributed parameter model is shown in Fig. A.1, which applies the RLC parameter of the line to an elemental section [75, 103].

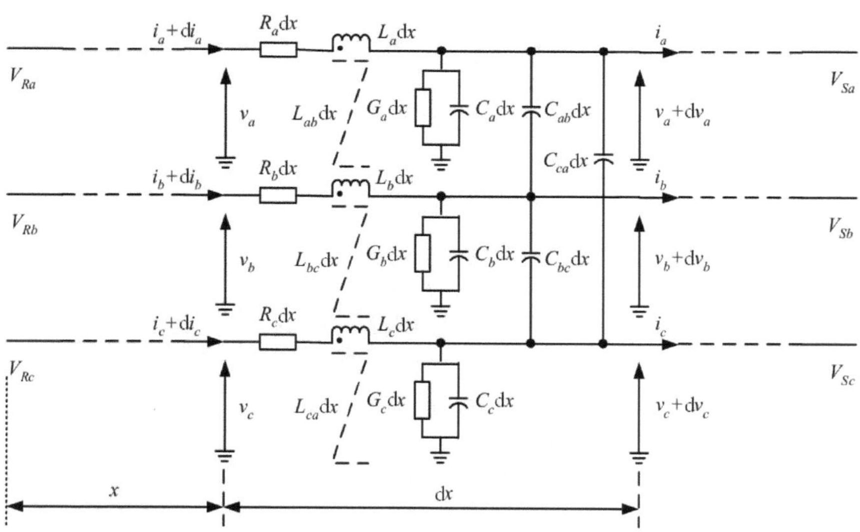

Fig. A.1 A distributed parameter circuit model for a three-phase transmission line

An elemental length dx of the three-phase line at a distance of x from a busbar at an instant t is considered. Assume that R_k, G_k, L_k, C_k are the self-resistance, conductance, inductance and capacitance per unit length of a kth phase, and L_{km}, C_{km} are the mutual inductance and capacitance per unit length between kth and mth phases, respectively, where k and m denote phase A, B and C. The voltage and current relationships of both ends of the elemental section are obtained by applying Kirchoff's law.

For phase A:

$$\mathrm{d}v_a = \frac{\partial v_a}{\partial x}\mathrm{d}x = \left(R_a\mathrm{d}x + L_a\mathrm{d}x\frac{\partial}{\partial t}\right) i_a + L_{ab}\mathrm{d}x\frac{\partial i_b}{\partial t} + L_{ca}\mathrm{d}x\frac{\partial i_c}{\partial t},$$
$$\mathrm{d}i_a = \frac{\partial i_a}{\partial x}\mathrm{d}x = \left(G_a\mathrm{d}x + C_a\mathrm{d}x\frac{\partial}{\partial t}\right) v_a + C_{ab}\mathrm{d}x\frac{\partial v_{ab}}{\partial t} + C_{ca}\mathrm{d}x\frac{\partial v_{ac}}{\partial t},$$

which is then simplified as:

$$\frac{\partial v_a}{\partial x} = \left(R_a + L_a\frac{\partial}{\partial t}\right) i_a + L_{ab}\frac{\partial i_b}{\partial t} + L_{ca}\frac{\partial i_c}{\partial t},$$
$$\frac{\partial i_a}{\partial x} = \left(G_a + C_a\frac{\partial}{\partial t}\right) v_a + C_{ab}\frac{\partial v_{ab}}{\partial t} + C_{ca}\frac{\partial v_{ac}}{\partial t}.$$

Similar equations can be derived for phases B and C. Then by replacing v_{ab}, v_{ac}, v_{ba}, v_{bc}, v_{ca}, v_{cb}, with their equivalent phase voltages, i.e. $v_{ab} = v_a - v_b, \ldots$, etc., and the differentiation operator $\frac{\partial}{\partial t}$ with the transform operator p, the three-phase current and voltage relationships can be written as:

$$\frac{\partial v}{\partial x} = Zi, \qquad \frac{\partial i}{\partial x} = Yv, \tag{A.1}$$

where

$$v = \begin{bmatrix} v_a \\ v_b \\ v_c \end{bmatrix}, \qquad i = \begin{bmatrix} i_a \\ i_b \\ i_c \end{bmatrix},$$

and

$$Z = \begin{bmatrix} R_a + L_a p & L_{ab}p & L_{ca}p \\ L_{ab}p & R_b + L_b p & L_{bc}p \\ L_{ca}p & L_{bc}p & R_c + L_c p, \end{bmatrix},$$

$$Y = \begin{bmatrix} G_a + C_{aa}p & C_{ab}p & C_{ca}p \\ C_{ab}p & G_b + C_{bb}p & C_{bc}p \\ C_{ca}p & C_{bc}p & G_b + C_{cc}p, \end{bmatrix},$$

where

$$C_{aa} = C_a + C_{ab} + C_{ca},$$
$$C_{bb} = C_{ab} + C_b + C_{bc},$$
$$C_{cc} = C_{ca} + C_{bc} + C_c.$$

By combining the two equations of (A.1), we have:

$$\frac{\partial^2 v}{\partial x^2} = Pv, \qquad \frac{\partial^2 i}{\partial x^2} = P^T i, \tag{A.2}$$

where

$$P = ZY, \qquad P^T = YZ.$$

Equation (A.2) is particularly useful for consideration of the important special cases of transmission lines.

A.2 Transient Wave Propagation Characteristics

Power transmission lines are normally of the three-phase type. However, it is much simpler to understand travelling wave concepts by first considering transient wave propagation in a single-phase line. The equations describing a single-phase line can be obtained directly from (A.1) by taking their scalar version:

$$\frac{\partial v}{\partial x} = Zi, \qquad \frac{\partial i}{\partial x} = Yv, \tag{A.3}$$

where v and i are the voltage and line current, respectively, measured at any point on the line; $Z = R + Lp$ and $Y = G + Cp$. In this case, the previously defined line parameters R, G, L, C represent single-phase line parameters per unit length.

When the two equations of (A.3) are combined, the scalar form of (A.2) is produced:

$$\frac{\partial^2 v}{\partial x^2} = \gamma^2 v, \qquad \frac{\partial^2 i}{\partial x^2} = \gamma^2 i, \tag{A.4}$$

where γ, known as the propagation constant, is a complex expression given by:

$$\gamma = \alpha + j\beta = \sqrt{ZY} = \sqrt{(R+Lp)(G+Cp)}.$$

The real part α is known as the attenuation constant, and the imaginary part β is known as the phase constant; β is measured in radian per unit length.

By solving (A.4) as an ordinary differential equation in x, we obtain:

$$v(x) = k_1 e^{-\gamma x} + k_2 e^{+\gamma x},$$
$$i(x) = \frac{1}{\sqrt{Z/Y}}(k_1 e^{-\gamma x} - k_2 e^{+\gamma x}),$$

where k_1 and k_2 are constants.

These equations are valid only at some fixed time, i.e. it is assumed that $v(x)$ and $i(x)$ are functions of the distance x along the line. However, since the line voltage and current are functions of the time t as well as the distance x, the equations need to be modified by replacing $v(x)$ and $i(x)$ with $v(x,t)$ and $i(x,t)$, and the constants k_1 and k_2 by the time functions $F_1(t)$ and $F_2(t)$, such that:

$$v(x,t) = e^{-\gamma x}F_1(t) + e^{+\gamma x}F_2(t),$$
$$i(x,t) = \frac{1}{\sqrt{Z/Y}}(e^{-\gamma x}F_1(t) - e^{+\gamma x}F_2(t)).$$

Assuming a lossless line, $R = G = 0$, which in turn leads to:

$$\sqrt{ZY} = p\sqrt{LC} = \frac{p}{c} = \gamma$$

and

$$\sqrt{\frac{Z}{Y}} = \sqrt{\frac{L}{C}} = Z_0,$$

then

$$v(x,t) = e^{-xp/c}F_1(t) + e^{+xp/c}F_2(t),$$
$$i(x,t) = (e^{-xp/c}F_1(t) - e^{+xp/c}F_2(t))/Z_0, \tag{A.5}$$

where

$$c = \frac{1}{\sqrt{LC}} \tag{A.6}$$

is known as the propagation speed and

$$Z_0 = \sqrt{\frac{L}{C}} \tag{A.7}$$

is known as the surge (characteristic) impedance.

It is possible, with the help of Taylor's theorem, to show that:

$$e^{\pm ap}f(t) = f(t \pm a).$$

If the relation is applied to (A.5), it can be obtained that:

$$\Delta v(x,t) = F_1\left(t - \frac{x}{c}\right) + F_2\left(t + \frac{x}{c}\right), \tag{A.8}$$
$$\Delta i(x,t) = \left(F_1\left(t - \frac{x}{c}\right) - F_2\left(t + \frac{x}{c}\right)\right)/Z_0. \tag{A.9}$$

Functions $F_1(t - x/c)$ and $F_2(t + x/c)$ thus represent travelling waves in the forward and backward directions of x, respectively, and (A.8) and (A.9) may therefore be written in the following simple form:

$$v = v^+ + v^-, \qquad i = i^+ + i^-, \tag{A.10}$$

where $v^+ = F_1(t - x/c)$ and $v^- = F_2(t + x/c)$ are the forward and backward voltage components, respectively. i^+ and i^- are similarly the forward and backward components of the current.

As can be seen from (A.8)–(A.10), the forward and backward components are related to each other by the surge impedance of the line as follows:

$$v^+ = Z_0 i^+, \qquad v^- = -Z_0 i^-. \tag{A.11}$$

It can be noted from the above analysis that the surge impedance Z_0 is a real value for an assumed lossless line and it is evident from (A.11) that the current components are simply a replica of their corresponding voltages. It also shows that, while the forward voltage and current waveforms are of the same sign, the backward voltage and current waveforms are of opposite sign.

A.3 Incidence, Reflection and Refraction of Transients

Transients travelling over assumed homogeneous lossless lengths of transmission line continue to propagate at a uniform speed c and are unchanged in shape. However, at points of discontinuity, such as open circuits or other line terminations, part of an incident wave is reflected back along the line and the rest is transmitted into and beyond the discontinuity. The wave impinging on the discontinuity is often called an incident wave, and the two waves to which it gives rise are normally referred to as reflected and transmitted waves.

Consider Fig. A.2. A line consists of two sections with different equivalent surge impedances, i.e. Z_{0RF} and Z_{0SF}, viewed from the junction F towards the line terminals R and S, respectively. Given that a transient travels along the line until it reaches point F, where only part of the wave passes to the section FS, and the rest is reflected back. Since the voltage or current of the transient at junction F must be continuous, the voltage or current a little to the left of junction F must be the same as those a little to the right. Thus, the following identical equations at any junction in the line should be satisfied:

$$\begin{aligned} v &= v^+ + v^- = v^{\mathrm{F}} + v^{\mathrm{B}} = v^{\mathrm{T}}, \\ i &= i^+ + i^- = i^{\mathrm{F}} + i^{\mathrm{B}} = i^{\mathrm{T}}, \end{aligned} \tag{A.12}$$

where superscripts F, B and T denote an incident (forward), a reflected (backward) and a transmitted wave. By applying (A.11), we obtain:

$$\begin{aligned} v^{\mathrm{F}} &= Z_{0\mathrm{RF}} i^{\mathrm{F}}, \\ v^{\mathrm{B}} &= -Z_{0\mathrm{RF}} i^{\mathrm{B}}, \\ v^{\mathrm{T}} &= Z_{0\mathrm{SF}} i^{\mathrm{T}}. \end{aligned} \tag{A.13}$$

Combining (A.12) and (A.13) results in:

$$v^{\mathrm{B}} = \frac{Z_{0\mathrm{SF}} - Z_{0\mathrm{RF}}}{Z_{0\mathrm{SF}} + Z_{0\mathrm{RF}}} v^{\mathrm{F}} = K_{\mathrm{R}} v^{\mathrm{F}}, \tag{A.14}$$

$$v^{\mathrm{T}} = \frac{2Z_{0\mathrm{SF}}}{Z_{0\mathrm{SF}} + Z_{0\mathrm{RF}}} v^{\mathrm{F}} = K_{\mathrm{T}} v^{\mathrm{F}}, \tag{A.15}$$

where constants K_{R} and K_{T} are known as the reflection and refraction coefficients, respectively.

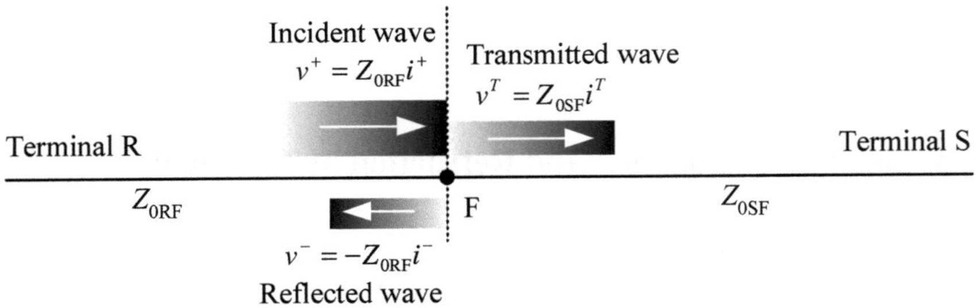

Fig. A.2 Schematic illustration of the incident, reflected and transmitted transient waves

References

1. Alternative Transients Program Rule Book. Leuven EMTP Centre, Heverlee, Belgium (1987)
2. Summary of IEEE guide for protection of network transformers. IEEE Transactions on Power Delivery **15**, 1496–1505 (1990)
3. Aggarwal, R.K., Johns, A.T., Song, Y.H., Dum, R.W., Fitton, D.S.: Neural-network based adaptive single-pole autoreclosure technique for EHV transmission systems. IEE Proceedings: Generation, Transmission and Distribution **141**(2), 155–160 (1994)
4. Aggarwal, R.K., Song, Y.H., Johns, A.T.: Adaptive three-phase autoreclosure for double-circuit transmission systems using neural networks. In: IEE 2nd International Conference on Advances in Power System Control, Operation and Management, pp. 389–392. Hong Kong (1993)
5. Aggarwal, R.K., Xuan, Q.Y., Dunn, R.W., Johns, A.T., Bennett, A.: A novel fault classification technique for double-circuit lines based on a combined unsupervised/supervised neural network. IEEE Transactions on Power Delivery **14**(4), 1250–1256 (1999)
6. Aguilera, C., Orduna, E., Ratta, G.: Adaptive noncommunication protection based on traveling waves and impedance relay. IEEE Transactions on Power Delivery **21**(3), 1154–1162 (2006)
7. AlFuhaid, A.S., El-Sayed, M.A.: A recursive least-squares digital distance relaying algorithm. IEEE Transactions on Power Delivery **14**(4), 1257–1262 (1999)
8. ALSTOM T&D Control Ltd.: Type MBCH Biased Differential Relay (1999). Service Manual
9. Anderson, P.M.: Power System Protection. McGraw-Hill, IEEE Press, The Institute of Electrical and Electronics Engineers Inc., New York (1998)
10. Annakkage, U.D., McLaren, P.G., Dirks, E., Jayasinghe, R.P., Parker, A.D.: A current transformer model based on the Jiles Atherton theory of ferromagnetic hysteresis. IEEE Transactions on Power Delivery **15**(1), 57–61 (2000)
11. Bastard, P., Bertrand, P., Meunier, M.: A transformer model for winding fault studies. IEEE Transactions on Power Delivery **9**(2), 690–699 (1994)
12. Benmouyal, G.G.: Removal of DC-offset in current waveforms using digital mimic filtering. IEEE Transactions on Power Delivery **10**(2), 621–630 (1995)
13. Bertrand, P., Devalland, A., Bastard, P.: A simulation model for transformer internal faults base for the study of protection and monitoring systems. In: 12th International Conference on Electricity Distribution, vol. 1, pp. 1.21/1–1.21/5. Birmingham, U.K. (1993)
14. Bewley, L.V.: Travelling waves on transmission systems. Wiley, New York (1951). Second edition
15. Bhalja, B.R., Maheshwari, R.P.: High-resistance faults on two terminal parallel transmission line: Analysis, simulation studies, and an adaptive distance relaying scheme. IEEE Transactions on Power Delivery **22**(2), 801–812 (2007)
16. Bickford, J.P., Abdel-Rahman, M.H.: Application of travelling-wave methods to the calculation of transient-fault currents and voltages in power-system networks. IEE Proceedings Part C **127**(3), 153–168 (1980)

17. Bloch, I., Maitre, H.: Fuzzy mathematical morphologies: A comparative study. Pattern Recognition **28**(9), 1341–1387 (1995)
18. Blume, L.F.: Transformer Engineering. Wiley, New York (1951)
19. Bo, Z.Q.: A new non-communication protection technique for transmission lines. IEEE Transactions on Power Delivery **13**(4), 1073–1078 (1998)
20. Bo, Z.Q., Aggarwal, R.K., Johns, A.T., Li, H.Y., Song, Y.H.: A new phase selection technique based on fault generated high frequency noise using neural networks. IEEE Transactions on Power Delivery **12**(1), 106–115 (1997)
21. Bo, Z.Q., Weller, G., Lomas, T.: A new technique for transformer protection based on transient detection. IEEE Transactions on Power Delivery **15**(3), 870–875 (2000)
22. Breingan, W.D., Chen, M.M.: The laboratory investigation of a digital system for the protection of transmission lines. IEEE Transactions on Power Apparatus and Systems **98**(2), 550–568 (1979)
23. Bunyagul, T., Crossley, P., Gale, P.: Overcurrent protection using signals derived from saturated measurement CTs. In: IEEE Power Engineering Society Summer Meeting, vol. 1, pp. 103–108. Vancouver, BC, Canada (2001)
24. Chamia, M., Liberman, S.: Ultra high speed relay for EHV/UHV transmission lines – development, design and application. IEEE Transactions on Power Apparatus and Systems **PAS-97**(6), 2104–2116 (1978)
25. Chen, C.S., Liu, C.W., Jiang, J.A.: Application of combined adaptive fourier filtering technique and fault detector to fast distance protection. IEEE Transactions on Power Delivery **21**(2), 619–626 (2006)
26. Chen, T., Wu, Q.H., Rahmani-Torkaman, R., Hughes, J.: A pseudo top-hat mathematical morphological approach to edge detection in dark regions. Pattern Recognition **35**(1), 199–210 (2002)
27. Chen, W., Malik, O.P., Yin, X.G., Chen, D.S., Zhang, Z.: Study of wavelet-based ultra high speed directional transmission line protection. IEEE Transactions on Power Delivery **18**(4), 1134–1139 (2003)
28. Christopoulos, C., Wright, A.: Electrical Power System Protection. Kluwer, Norwell, MA (1999). Second edition
29. Clarke, E.: Circuit Analysis of a.c. Power Systems. Wiley, New York (1943)
30. Claypoole, R., David, G., Sweldens, W., Baraniuk, R.: Nonlinear wavelet transforms for image coding. In: Conference Record of the 31st Asilomar Conference on Signals, Systems and Computers, vol. 1, pp. 662–667 (1997)
31. Claypoole, R.L., Baraniuk, R.G., Nowak, R.D.: Adaptive wavelet transforms via lifting. In: Proceedings of the 1998 IEEE International Conference on Acoustics, Speech and Signal Processing, vol. 3, pp. 1513–1516. Seattle, WA (1998)
32. Claypoole, R.L., Baraniuk, R.G., Nowak, R.D.: Lifting construction of nonlinear wavelet transforms. In: Proceedings of IEEE-SP International Symposium on Time-Frequency Time-Scale Analysis, pp. 49–52. Pittsburgh, PA (1998)
33. Claypoole, R.L., Baraniuk, R.G., Nowak, R.D.: Adaptive wavelet transforms via lifting. Technique Report 9304, Department of Electrical and Computer Engineering, Rice University, Houston, TX (1999)
34. Coltuc, D., Pitas, I.: Morphological residual representations of signals. IEEE Transactions on Image Processing **5**(11), 1569–1572 (1996)
35. Damborg, M.J., Kim, M., Huang, J., Venkata, S.S., Phadke, A.G.: Adaptive protection as preventive and emergency control. IEEE Power Engineering Society Summer Meeting **2**, 1208–1212 (2000)
36. Dmitrenko, A.M.: Semiconductor pulse-duration differential restraint relay. Radiophysics and Quantum Electronics (3), 335–339 (1970)
37. Dong, X.Z., Ge, Y.Z., He, J.L., Guo, X.J., Bo, Z.Q.: Status quo and prospect of traveling waves protection of transmission lines. Automation of Electric Power Systems **24**(5), 56–61 (2000)
38. Duhamel, P., Vetterli, M.: Fast Fourier transforms: A tutorial review and a state of the art. Signal Processing **19**(4), 259–299 (1990)

39. Eissa, M.M.: Ground distance relay compensation based on fault resistance calculation. IEEE Transactions on Power Delivery **21**(4), 1830–1835 (2006)
40. Eissa, M.M., Masoud, M.: A novel digital distance relaying technique for transmission line protection. IEEE Transactions on Power Delivery **16**(3), 380–384 (2001)
41. Elmore, W.A.: Protective Relaying. Theory and Applications. Marcel Dekker, New York (1994)
42. Fernández, C.: An impedance-based CT saturation detection algorithm for bus-bar differential protection. IEEE Transactions on Power Delivery **16**(7), 468–472 (2001)
43. Fernández, C., Pagola, F.L.: Total least squares and discrete-time line models in HV distance protection. IEEE Transactions on Power Delivery **4**(1), 74–79 (1999)
44. Franco, S.: Electric Circuits Fundamentals. Saunders College Publishing, Fort Worth, TX (1995)
45. Gallen, T.F., Breingan, W.D.: A digital system for directional comparison relaying. IEEE Transactions on Power Apparatus and Systems **98**(3), 948–956 (1979)
46. Gaouda, A.M., Salama, M.M.A.: Power quality detection and classification using wavelet multi-resolution signal decomposition. IEEE Transactions on Power Delivery **14**(4), 1469–1475 (1999)
47. Gasteratos, A., Andreadis, I.: Soft mathematical morphology: Extensions, algorithms and implementations. Advances in Imaging and Electron Physics **110**, 63–99 (1999)
48. Gasteratos, A., Andreadis, I., Tsalides, P.: Fuzzy soft mathematical morphology. IEE Proceedings: Vision, Image Signal Processing **145**(1), 41–49 (1988)
49. Gasteratos, A., Andreadis, I., Tsalides, P.: Realisation of soft morphological filters. IEE Proceedings: Circuits Devices and Systems **145**(3), 201–206 (1998)
50. Gilbert, J.G., Shovlin, R.J.: High speed transmission line fault impedance calculation using a dedicated minicomputer. IEEE Transactions on Power Apparatus and Systems **94**(3), 544–550 (1975)
51. Girgis, A.A.: A new Kalman filtering based digital distance relay. IEEE Transactions on Power Apparatus and Systems **101**(9), 3471–3480 (1982)
52. Girgis, A.A., Brown, R.G.: Application of Kalman filtering in computer relaying. IEEE Transactions on Power Apparatus and Systems **100**(7), 3387–3397 (1981)
53. Giuliante, T., Clough, G.: Advances in the design of differential protection of power transformers. In: Texas A&M University Conference for Protective Relay Engineers. College Station, TX (1995)
54. Gomez-Morante, M., Nicoletti, D.W.: A wavelet-based differential transformer protection. IEEE Transactions on Power Delivery **14**(4), 1351–1358 (1999)
55. Goutsias, J., Heijmans, H.J.A.M.: Nonlinear multiresolution signal decomposition schemes – Part I: Morphological pyramids. IEEE Transactions on Image Processing **9**(11), 1862–1876 (2000)
56. Gu, J.C., Yu, S.L.: Removal of DC offset in current and voltage signals using a novel Fourier filter algorithm. IEEE Transactions on Power Delivery **15**(1), 73–79 (2000)
57. Hase, Y.: Handbook of Power System Engineering. Wiley, West Sussex, U.K. (2007)
58. He, B.T., Li, Y.Q., Bo, Z.Q.: An adaptive distance relay based on transient error estimation of CVT. IEEE Transactions on Power Delivery **21**(4), 1856–1861 (2006)
59. Heideman, M.T., Johnson, D.H., Burrus, C.S.: Gauss and the history of the fast Fourier transform. IEEE Acoustics Speech and Signal Processing Magazine **1**(4), 14–21 (1984)
60. Heijmans, H.J.A.M., Goutsias, J.: Nonlinear multiresolution signal decomposition schemes – Part II: Morphological wavelets. IEEE Transactions on Image Processing **9**(11), 1897–1913 (2000)
61. Horowitz, S.H., Phadke, A.G.: Power System Relaying. Wiley, New York (1992)
62. Horton, J.W.: The use of walsh functions for high-speed digital relaying. In: IEEE PES Summer Meeting, pp. 582–587. San Francisco, CA (1975)
63. Hu, Y., Novosel, D., Saha, M.M., Leitloff, V.: An adaptive scheme for parallel-line distance protection. IEEE Transactions on Power Delivery **17**(1), 105–110 (2002)
64. Huang, S.J., Hsieh, C.T.: High-impedance fault detection utilizing a morlet wavelet transform approach. IEEE Transactions on Power Delivery **14**(4), 1401–1410 (1999)

65. Izykowski, J., Rosolowski, E., Saha, M.M.: Locating faults in parallel transmission lines under availability of complete measurements at one end. IEE Proceedings: Generation, Transmission and Distribution **151**(2), 268–273 (2004)
66. Ji, T.Y., Lu, Z., Wu, Q.H.: Optimal soft morphological filter for periodic noise removal using a particle swarm optimiser with passive congregation. Signal Processing **87**(11), 2799–2809 (2007)
67. Ji, T.Y., Lu, Z., Wu, Q.H.: Baseline normalization of ECG signals using empirical mode decomposition and mathematical morphology. Electronics Letters **44**(2), 82–83 (2008)
68. Ji, T.Y., Lu, Z., Wu, Q.H.: Detection of power disturbances using morphological gradient wavelet. Signal Processing **88**(2), 255–267 (2008)
69. Ji, Z., Li, H.H., Li, Q., Wu, Q.H.: Fault location in power transmission line system using adaptive lifting scheme. Electronics Letters **41**(7), 440–441 (2005)
70. Johns, A.T.: New ultra-high-speed directional comparison technique for the protection of EHV transmission lines. IEE Proceedings Part C **127**(4), 228–239 (1980)
71. Johns, A.T., Aggarwal, R.K.: Digital simulation of faulted EHV transmission lines with particular reference to very-high-speed. IEE Procedings Part C **123**(4), 354–359 (1976)
72. Johns, A.T., Aggarwal, R.K., Bo, Z.Q.: Non-unit protection technique for EHV transmission systems based on fault-generated noise, Part 1: Signal measurement. IEE Proceedings on Generation, Transmission and Distribution **141**(2), 133–140 (1994)
73. Johns, A.T., Aggarwal, R.K., Song, Y.H.: Improved techniques for modelling fault arcs on faulted EHV transmission systems. IEE Proceedings on Generation, Transmission and Distribution **141**(2), 148–154 (1994)
74. Johns, A.T., Martin, M.A.: Fundamental digital approach to the distance protection of EHV transmission lines. IEE Proceedings **125**(5), 377–384 (1978)
75. Johns, A.T., Salman, S.K.: Digital Protection for Power Systems. Institution of Electrical Engineers, London (1995)
76. Jones, M.S., Thomas, D.W.P., Christopoulos, C.: A non-pilot phase selector based on superimposed components for protection of double circuit lines. IEEE Transactions on Power Delivery **12**(4), 1439–1444 (1997)
77. Kang, Y.C., Ok, S.H., Kang, S.H.: A CT saturation detection algorithm. IEEE Transactions on Power Delivery **19**(1), 78–85 (2004)
78. Kang, Y.C., Park, J.K., Kang, S.H., Johns, A.T., Aggarwal, R.K.: An algorithm for compensating the secondary current of current transformers. IEEE Transactions on Power Delivery **12**(1), 116–124 (1997)
79. Karsai, K., Kerenyi, D., Kiss, L.: Large Power Transformers. Elsevier, Amsterdam (1987)
80. Kasztenny, B., Kezunovic, M.: Improved power transformer protection using numerical relays. IEEE Computer Applications in Power **11**(4), 39–45 (1998)
81. Kasztenny, B., Kulidjian, A.: An improved transformer inrush restraint algorithm increases security while maintaining fault response performance. In: 5th Annual Conference for Protective Relay Engineers. College Station, TX (2000)
82. Kasztenny, B., Rosolowski, E., Saha, M.M., Hillstrom, B.: A comparative analysis of protection principles for multi-citeria power transformer relaying. In: Proceedings of the 12th Power Systems Computation Conference, pp. 107–113. Dresden, Germany (1996)
83. Koskinen, L., Astola, J., Neuvo, Y.: Soft morphological filters. In: Proceedings of SPIE Symposium on Image Algebra and Morphological Image Processing, pp. 262–270 (1991)
84. Kreyszig, E.: Advanced Engineering Mathmatics. Wiley, New York (1995)
85. Kulidjian, A., Kasztenny, B., Campbell, B.: New magnetizing inrush restraining algorithm for power transformer protection. In: Proceedings of the 7th International Conference on Developments in Power System Protection, pp. 181–184. Amsterdam (2001)
86. Kuosmanen, P., Astola, J.: Soft morphological filtering. Mathematical Imaging and Visions **5**(3), 231–262 (1995)
87. Lang, M., Guo, H., E.Odegard, J., Burrus, C.S., Wells, R.O.: Noise reduction using an undecimated discrete wavelet transform. Signal Processing Letters **3**(1), 10–12 (1996)
88. Lewis, L.J.: Travelling wave relations applicable to power-system fault locators. AIEE Transactions **50**, 1671–1680 (1951)

89. Liu, P., Malik, O.P., Chen, C., Hope, G.S., Guo, Y.: Improved operation of differential protection of power transformers for internal faults. IEEE Transactions on Power Delivery **7**(4), 1912–1919 (1992)
90. Lu, Z., Ji, T.Y., Wu, Q.H.: Morphological transform for removal of exponentially decaying DC-offset. Electronics Letters **44**(9), 595–596 (2008)
91. Lu, Z., Smith, J.S., Wu, Q.H.: Morphological lifting scheme for current transformer saturation detection and compensation. IEEE Transactions on Circuits and Systems I: Regular Papers **55**(10), 3349–3357 (2008)
92. Lu, Z., Tang, W.H., Ji, T.Y., Jiang, L., Wu, Q.H.: A phase selector based on mathematical morphology for double circuit transmission lines. In: Third International Conference on Electric Utility Deregulation and Restructuring and Power Technologies, DRPT 2008, 9, pp. 2322–2327. Nanjing, China (2008)
93. Lu, Z., Tang, W.H., Ji, T.Y., Wu, Q.H.: A morphological scheme for inrush identification in transformer protection. IEEE Transactions on Power Delivery. In press (2009)
94. Lucas, J.R.: Representation of magnetization curves over a wide region using non-integer power series. International Journal of Electrical Engineering Education **25**(4), 335–340 (1988)
95. Magnago, F.H., Abur, A.: Fault location using wavelets. IEEE Transactions on Power Delivery **13**(4), 1475–1480 (1998)
96. Mallat, S.: A theory for multiresolution signal decomposition: The wavelet representation. IEEE Transactions on Pattern Analysis and Machine Intelligence **11**(7), 674–693 (1989)
97. Mallat, S.: A Wavelet Tour of Signal Processing. Academic Press, San Diego, CA (1999)
98. Mann, B.J., Morrison, I.F.: Digital calculation of impedance for transmission line protection. IEEE Transactions on Power Apparatus and Systems **90**(1), 270–279 (1971)
99. Matheron, G.: Random sets and integrated geometry. Wiley, New York (1975)
100. Matheron, G.: Elments pour une theorie des milieux poreux. Masson, Paris (1976)
101. McLaren, P.G., Redfern, M.A.: Fourier-series techniques applied to distance protection. IEE Proceedings **122**(11), 1301–1305 (1975)
102. Moravej, Z., Vishwakarma, D.N., Singh, S.P.: ANN-based protection scheme for power transformer. Electric Machines and Power Systems **28**(9), 875–884 (2000)
103. Ong, C.M.: Dynamic Simulation of Electric Machinery Using MATLAB/SIMULINK. Prentice Hall, Englewood Cliffs, NJ (1997)
104. Osman, A.H., Malik, O.P.: Transmission line distance protection based on wavelet transform. IEEE Transactions on Power Delivery **19**(2), 515–523 (2004)
105. Pan, J.P., Vu, K., Hu, Y.: An efficient compensation algorithm for current transformer saturation effects. IEEE Transactions on Power Delivery **19**(4), 1623–1628 (2004)
106. Pathirana, V., McLaren, P.G.: A hybrid algorithm for high speed transmission line protection. IEEE Transactions on Power Delivery **20**(4), 2422–2428 (2005)
107. Perez, L.G., Flechsig, A.J., Meador, J.L., Obradoviic, Z.: Training an artificial neural network to discriminate between magnetizing inrush and internal faults. IEEE Transactions on Power Delivery **9**(1), 434–441 (1994)
108. Phadke, A.G., Ibrahim, M., Hlibka, T.: Fundamental basis for distance relaying with symmetrical components. IEEE Transactions on Power Apparatus and Systems **PAS-96**(2), 635–646 (1977)
109. Phadke, A.G., Thorp, J.S.: Computer Relaying for Power Systems. Research Studies Press, New York (1988)
110. Pihler, J., Grcar, B., Dolinar, D.: Improved operation of power transformer protection using artifical neural network. IEEE Transactions on Power Delivery **12**(3), 1128–1136 (1997)
111. Pitas, I., Venetsanopoulos, A.N.: Morphological shape decomposition. IEEE Transactions on Pattern Analysis and Machine Intelligence **12**(1), 38–45 (1990)
112. Radojević, Z.M., Shin, J.R.: New one terminal digital algorithm for adaptive reclosing and fault distance calculation on transmission lines. IEEE Transactions on Power Delivery **21**(3), 1231–1237 (2006)

113. Ranjbar, A.M., Cory, B.J.: An improved method for the digital protection of high voltage transmission lines. IEEE Transactions on Power Apparatus and Systems **92**(2), 544–550 (1975)
114. Rockefeller, G.D.: Fault protection with digital computer. IEEE Transactions on Power Apparatus and Systems **88**(4), 438–461 (1969)
115. Rockefeller, G.D., Gilchrist, G.B.: High-speed distance relaying using a digital computer, II. Test results. IEEE Transactions on Power Apparatus and Systems **91**(3), 1244–1258 (1972)
116. Rockefeller, G.D., Wagner, C.L., Linders, J.R., Hicks, K.L., Rizy, D.T.: Adaptive transmission system relaying concepts for improved performance. IEEE Transactions on Power Delivery **3**(4), 1146–1458 (1988)
117. Russel, B.D., Council, M.E.: Power System Control and Protection. Academic Press, New York (1978)
118. Sachdev, M.S.: Computer Relaying – Tutorial Course Text. IEEE Publications No. 79EH0148-7-PWR, New York (1979)
119. Sachdev, M.S., Baribeau, M.A.: A new algorithm for digital impedance relays. IEEE Transactions on Power Apparatus and Systems **98**(6), 2232–2240 (1979)
120. Saleh, S.A., Rahman, M.A.: Modeling and protection of a three-phase power transformer using wavelet packet transform. IEEE Transactions on Power Delivery **20**(2), 1273–1282 (2005)
121. Samantaray, S.R., Dash, P.K.: Transmission line distance relaying using a variable window short-time Fourier transform. Electric Power Systems Research **78**(4), 595–604 (2008)
122. Sedaaghi, M.H., Wu, Q.H.: Real-time implementation of grey-scale morphological operators. Electronics Letters **33**(21), 1761–1763 (1997)
123. Sedaaghi, M.H., Wu, Q.H.: Weighted morphological filter. Electronics Letters **34**(16), 1566–1567 (1998)
124. Serra, J.: Automatic scanning device for analysing textures. Patent, Belgium, Canada, Great-Britain, Japan, Sweden, U.S. **23**(273) (1965)
125. Serra, J.: Image analysis and mathematical morphology. Academic Press, London (1982)
126. Serra, J.: Image Analysis and Mathematical Morphology II: Theoretical Advances. Academic Press, London (1988)
127. Shinha, D., Dougherty, E.R.: Fuzzy mathematical morphology. Visual Communication and Image Representation **3**(3), 286–302 (1992)
128. Sidhu, T.S., Ghotra, D.S., Sachdev, M.S.: An adaptive distance relay and its performance comparison with a fixed data window distance relay. IEEE Transactions on Power Delivery **17**(3), 691–697 (2002)
129. Sidhu, T.S., Sachdev, M.S., Wood, H.C., Nagpal, M.: Design, implementation and testing of a micor-processor-based high speed relay for detecting transformer winding faults. IEEE Transactions on Power Delivery **7**(1), 108–117 (1992)
130. Smolinski, W.: An algorithm for digital impedance calculation using a single PI section transmission line model. IEEE Transactions on Power Apparatus and Systems **98**(5), 1546–1551 (1979)
131. Soille, P.: Morphological image analysis. Springer, Berlin (2003)
132. Sternberg, S.: Grayscale morphology. Computer Vision, Graphics, and Image Processing **35**(3), 333–355 (1986)
133. Stringer, N.T.: The effect of DC offset on current-operated relays. IEEE Transactions on Industry Applications **34**(1), 30–34 (1998)
134. Sun, P., Wu, Q.H., Weindling, A.M., Finkelstein, A., Ibrahim, K.: An improved morphological approach to background normalization of ECG signals. IEEE Transactions on Biomedical Engineering
135. Sun, P., Zhang, J.F., Zhang, D.J., Wu, Q.H.: Morphological indentification of transformer magnetising inrush current. IEE Electronics Letters **38**(9), 437–438 (2002)
136. Sweldens, W.: Wavelets and the lifting scheme: A 5 minute tour. Zeitschrift für Angewandte Mathematik und Mechanik **76**, 41–44 (1996)
137. Sweldens, W.: The lifting scheme: A construction of second generation wavelets. SIAM Journal on Mathematical Analysis **29**(2), 511–546 (1998)

138. Sykes, J.A., Morrison, I.F.: A proposed method of harmonic restraint differential protection of transformers by digital computer. IEEE Transactions on Power Apparatus and Systems **PAS-91**(3), 1266–1272 (1972)
139. Thorp, J.S., Phadke, A.G., Horowitz, S.H., Begovic, M.M.: Some applications of phasor measurements to adaptive protection. IEEE Transactions on Power Delivery **3**(2), 197–798 (1988)
140. van den Boomgaard, R.: Mathematical morphology: Extensions towards computer vision. Ph.D. thesis, University of Amsterdam (1992)
141. Vitins, M.: A fundamental concept for high speed relaying. IEEE Transcations on Power Apparatus and Systems **100**(1), 163–168 (1981)
142. Waldron, J.E., Zocholl, S.E.: Design considerations for a new solid state transformer differential relay with harmonic restraint. In: Fifth Western Protective Relay Conference. Sacramento, CA (1978)
143. Wiszniewski, A., Kasztenny, B.: A multi-criteria differential transformer relay based on fuzzy logic. IEEE Transactions on Power Delivery **10**(4), 1786–1792 (1995)
144. Wu, Q.H., Sun, P., Zhang, D.J., Zhang, J.F.: Identification of transformer inrush current using morphological signal decomposition. Automation of Electric Power Systems **26**(Supplement), 67–78 (2002)
145. Wu, Q.H., Zhang, J.F., Zhang, D.J.: Ultra-high-speed directional protection of transmission lines using mathematical morphology. IEEE Transactions on Power Delivery **18**(4), 1127–1133 (2003)
146. Xuan, Q.Y., Aggarawal, R.K., Johns, A.T., Dunn, R.W., Bennett, A.: A neural network based protection technique for combined 275 kV/400 kV double circuit transmission lines. International Journal of Neurocomputing **23**, 59–70 (1998)
147. Yalla, M.V.V.S.: A digital multifunction protective relay. IEEE Transactions on Power Delivery **7**(1), 193–201 (1992)
148. Youssef, O.A.S.: New algorithm to phase selection based on wavelet transforms. IEEE Transactions on Power Delivery **17**(4), 908–914 (2002)
149. Youssef, O.A.S.: A wavelet-based technique for discrimination between faults and magnetizing inrush currents in transformers. IEEE Transactions on Power Delivery **18**(1), 170–176 (2003)
150. Yu, D.C., Wang, Z., Cummins, J.C., Yoon, H.J., Kojovic, L.A., Stone, D.: Neural network for current transformer saturation correction. In: 1999 IEEE Transmission and Distribution Conference, vol. 1, pp. 441–446. New Orleans, LA (1999)
151. Zhang, D.J., Wu, Q.H.: Identification of magnetizing inrush current based on wavelet transform. Technique Report, Department of Electrical Engineering and Electronics, The University of Liverpool (2000)
152. Zhang, D.J., Wu, Q.H., Bo, Z.Q., Caunce, B.: Transient positional protection of transmission lines using complex wavelets analysis. IEEE Transactions on Power Delivery **18**(3), 705–710 (2003)
153. Zhang, D.J., Wu, Q.H., Zhang, J.F., Nuttall, K.I.: Accurate fault location based on transients extraction using mathematical morphology. Electronics Letters **38**(24), 1583–1585 (2002)
154. Zhang, J.F., Smith, J.S., Wu, Q.H.: Morphological undecimated wavelet decomposition for fault location on power transmission lines. IEEE Transactions on Circuits and Systems I: Regular Papers **53**(6), 1395–1402 (2006)
155. Zhang, Z.Z., Chen, D.S.: An adaptive approach in digital distance protection. IEEE Transactions on Power Delivery **11**(1), 44 (1991)
156. Ziegler, G.: Numerical Distance Protection: Principles and Applications. Wiley VCH, Berlin (2008). Third edition

Index

MIX
Papier aus verantwortungsvollen Quellen
Paper from responsible sources
FSC® C105338

If you have any concerns about our products,
you can contact us on
ProductSafety@springernature.com

In case Publisher is established outside the EU,
the EU authorized representative is:
Springer Nature Customer Service Center GmbH
Europaplatz 3, 69115 Heidelberg, Germany

Printed by Libri Plureos GmbH
in Hamburg, Germany